CURIOSITY
GUIDES

GLOBAL CLIMATE CHANGE

**CURIOSITY
GUIDES**

GLOBAL CLIMATE CHANGE

The Book of Essential Knowledge

Ernest Zebrowski, Ph.D.

Foreword by Alice Madden, LL.D.

imagine!
Publishing

An Imagine Book
Published by Charlesbridge
85 Main Street
Watertown, MA 02472
(617) 926-0329
www.charlesbridge.com

Library of Congress Cataloging-in-Publication Data
Zebrowski, Ernest
Curiosity guides : global climate change / Zebrowski, Ernest c.
    p. cm.
Includes bibliographical references and index.
ISBN 978-1-936140-16-9 (hardcover : alk. paper)
1. Global temperature changes. 2. Global warming. 3. Nature--Effect of
human beings on. I. Title.

(hc) 10 9 8 7 6 5 4 3 2 1

Illustrations created in Adobe Illustrator by Haude Levesque
Display type and text type set in Weiss, Garamond, and Granjon
Manufactured in China, October, 2010
Designed by Linda Kosarin / theartdepartment.biz

CONTENTS

ACKNOWLEDGMENTS

To those climate scientists whose work I have left out of this book, I apologize. There was a limit to what I could fit in, and I had to make some painful choices. Hopefully, I have chosen well enough to present a reasonably coherent and error-free narrative, and to stimulate some readers to follow future developments and discoveries on their own.

My special thanks to my dear friend and sometime collaborator Judith A. Howard, psychotherapist and political columnist, who waded through several drafts of this project and offered numerous perceptive comments. Thanks also to my niece Mariah Zebrowski, an environmentalist and soon-to-be environmental lawyer, who spotted a number of pre-publication errors and made dozens of other valuable suggestions. To David B. Clark, chemist, environmental scientist, and friend, my sincere appreciation for catching several scientific errors before they snuck into print. Also to Tim Ritter, physicist, colleague, and another friend—my gratitude for spending part of your vacation reviewing the manuscript; your input was invaluable.

Last, but hardly least, I thank my son, David Zebrowski, and my daughter, Angel Pruszenski, for taking time from their respective studies to act as sounding boards for my ideas; they, too, will notice their intellectual imprints in this book.

As for Jake, and despite the fact that he'll always have the mentality of a two-year-old, he is without doubt the best Labrador retriever on the island, and is a constant reminder to me that if we humans mess up the planet, our world will lose a lot more than humans.

ERNEST ZEBROWSKI
St. George Island, FL, USA

FOREWORD

I wish to thank Ernest Zebrowski for this vastly interesting historical journey through the discoveries that brought the world to its current understanding of global climate change. His interdisciplinary approach will capture the attention of a broad range of interests and provide a solid foundation of understanding from which readers can draw for years to come.

The last ten years of my life have been immersed in policy and politics, and I have learned that "winning the hearts and minds" of the people is not just bluster. Arming individuals with knowledge from trusted sources is the cornerstone of any voluntary change in behavior; and it is my belief that individual action will, ultimately, be the linchpin of our fate.

Even under the best of circumstances, motivating people to alter their habits is difficult. When the objective is stemming the consequences of climate change, it can be overwhelming—to the point of paralysis. In the economically advanced sectors and nations of the world—the United States, Western Europe, and elsewhere—one can add to the mix a concerted media campaign

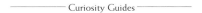

not only encouraging inaction, but painting those seeking change as extremist, alarmist, antibusiness, or even unpatriotic.

It is not too complicated to connect the dots from the loudest "deniers" to powerful industries, whose profits rely on maintaining the status quo. Yet, as long as detractors successfully distract and dither, ground is being lost in respect of leading- edge research—thereby jeopardizing the economic security of such recalcitrants, be they lobbyists, organizations, or entire governments.

This timely book offers us a way to deflect the barrage of self-serving misinformation. The next step, equally important, is building momentum to embrace the endless opportunities—and to understand the following important points.

- Investing in renewable energy research will trigger a manufacturing renaissance.
- Utilizing more natural gas will provide cleaner and more efficient energy.
- What we consider currently to be mere waste can be turned into energy.
- All such efforts help build a sustainable economy and improve our natural environment.

We know that saving money is a powerful agent of change, and many energy reduction programs offer incentives such as rebates and tax credits. But we still must get to the matters of the heart. What creates true believers—those who will become thought leaders of their communities? A growing number of religious groups are now incorporating climate change as part

of an ethos that includes being faithful stewards of the Earth. For others, the turning point may be as simple as an iconic photo of a stranded polar bear. For me, as an American, it was the devastation of our western forests, and all that they represent, by the bark beetle. Yet for others around the globe, it might take the unfathomable loss of entire island nations and their ancient cultures.

Future generations will, no doubt, be more in tune to these issues by sheer force of necessity. It is my very great hope that we, the world citizens of today, do everything we can to give them the head start that they deserve.

ALICE MADDEN, LL.D.

Former Majority Leader, Colorado State House of Representatives
Current Climate Change Advisor to the Governor of Colorado.

PREFACE

This is a story that might well be described as the most extensive forensic investigation ever conducted. It also surely ranks among the most difficult.

Thousands of investigators analyzed millions of shreds of data before it was even clear that there was a victim. Then, when the evidence started piling up that our whole biosphere as we know it is indeed threatened, it still was not obvious that there was a perpetrator. In fact, far from entertaining the prospect that we humans might be the actual culprits, many experts found it easy to ignore or pooh-pooh the initial claims that something dreadful was going on.

At first, the symptoms were called "global warming." That terminology had a couple of flaws. Firstly, there are a few places on the globe that do not seem to be warming, and secondly, changes in other climatic variables like precipitation patterns are potentially every bit as troublesome as rising temperatures. So numerous writers, myself included, have found the phrase "global climate change" to be more appropriate.

As to *why* Earth's climate is changing, the word "anthropogenic" keeps appearing in the scientific literature—human-generated, in other words. Climate, of course, can also be altered by nonhuman factors, and indeed that has happened numerous times in the distant past. But this time, something different seems to be going on. Human activities are now changing our planet's atmosphere and its hydrosphere (the oceans, lakes, rivers, etc.) in ways that almost certainly bode dire consequences for the future: rising sea levels, altered patterns of storms and droughts, and shifts in agricultural productivity, for instance.

It would be remarkable if the collective activities of 6.77 billion humans (the estimated world population at the time of writing) did not affect Earth's climate. But deciphering exactly what those effects are—and their implications for the future of the planet—are not simple tasks. Although the world of science occasionally has its big breakthroughs, for the most part it plods along slowly. There is a bit of data here, an analysis there, journal articles from time to time, objections and responses to the articles, further experiments and field studies, and so on. Meanwhile, there are always those individuals who cherry-pick or even concoct "facts" that will benefit their various self-interests. Indeed, some of these unprincipled characters have a great deal more financial and political clout than any mere scientist. As a result, the issue of global climate change has become mired in politics, particularly in the United States.

Political arguments, however, are not the focus of this book. Politicians and business leaders can pontificate about what they want to believe to be the scientific truths, but Mother Nature will always be the final arbiter. It is only through scientific inquiry that we humans can understand how our environment actually behaves. And it is only through scientific understanding that we may have some chance of remedying or reversing a looming global climate crisis.

Meanwhile, new climatological findings are published almost daily. Many are mere tweaks to prior information, while others venture further into the realm of speculation about our planet's future. I have taken a cautious approach to this bank of evolving scientific information: If an ongoing study has not been independently replicated, I chose to omit it regardless of how much I personally may be impressed or troubled by it. By the time this book reaches you, the reader, however, some of these more recent discoveries and/or refutations will surely have added to our overall scientific picture of global climate change.

So what *have* I included in this book? The fascinating tale of a century and a half of investigations that did and did not pan out, and which collectively settle any issue of reasonable doubt about the reality of global climate change. This book is the story of hundreds of researchers who sought to unravel the mysteries of Nature, and in so doing revealed a global process that threatens to transform our planet into a considerably less hospitable place.

What follows, then, is the story of the science—the *how*s and *why*s of our human efforts to probe the mind of Mother Nature, who, after all, is not subject to human political arguments.

CHAPTER 1

The Iceman and the Glaciers

You never know your friends from your enemies until the ice breaks.
ESKIMO PROVERB

On September 19, 1991, vacationers Helmut and Erika Simon made a distressing discovery while hiking in the Ötztal Alps near the Austrian-Italian border. Helmut was tramping along ahead of his wife in order to test the footing, just as he always did when the terrain became steep. He was also carrying his camera, just as he usually did on their outings. It had been a photogenic day, and he had just two exposures left.

AN ICEMAN COMETH

In the glacial ice of a nearby ravine, at an altitude of 10,500 feet (3200 m), Helmut noticed what appeared to be some litter discarded by a careless hiker. On closer inspection, a shiver ran up his spine. Poking out of the melting ice, facedown, was the upper third of a human corpse. Overruling Erika's objections about photographing a dead person, Helmut snapped one of their two remaining pictures.

On their return trek down the mountain, the Simons discussed the unnerving experience. They'd heard that every once in a while a hiker might die in an alpine blizzard, only to remain undiscovered for years. In fact, just three weeks earlier, a couple that had gone missing in 1934 had been found, their bodies having been interred in the alpine snow for 57 years. The corpse that Helmut and Erika had seen and photographed could possibly be that old; but, then again, it could also be much more recent.

If they reported their find to the authorities, there would most likely be an investigation that could ruin the remainder of their vacation. Not only that, but they might be delayed from returning to Germany until the matter was legally resolved. Maybe they should just keep quiet. Surely someone else would soon notice the body.

Their return trail intersected a narrow road, and there stood a rustic lodge. They stopped for a drink. As they sat sipping their beers and mulling over the matter, they agreed that, after all, they were not the kind who could let a human death go unreported. They related their experience to the innkeeper, Markus Pirpamer, and sketched him a map of the body's location. As Pirpamer phoned the Austrian authorities, the Simons continued down the mountain to their hotel. A few days later, they returned to Germany.

On September 21, an Austrian recovery team arrived to

retrieve the corpse, which they found problematically encased in ice below the torso. One worker grabbed a nearby stick to try to pry it loose. The wooden stick, which promptly splintered, would turn out to be a prehistoric hunting bow.

More mistakes were made. They tugged on the body's clothing, unwittingly shredding ancient hides and fibers. One team member took a small jackhammer to the ice and punched a hole in the deceased's hip. When the team finally exhumed the body, using manual ice picks, they broke its left arm when jamming it into a coffin.

During the corpse's next few days in a local morgue, a fungus spread over its thawing skin. Only gradually did it dawn on anyone that the dead man was a Bronze Age hunter. Radiocarbon dating would set his date of death at between 3350 and 3300 BCE.

Scientists named the unfortunate fellow Ötzi, after the Ötztal Alps where he had been discovered. In fact, Ötzi had been frozen solid for more than 5,000 years, only to reemerge from a melting glacier.

The cause of death posed a few problems. It was probably not exposure, or at least not exposure alone. Ötzi's shoes and leather clothing were remarkably sophisticated, skillfully crafted, and clearly designed to provide insulation against the cold. Starvation was also ruled out, because he had had a meal of wheat, vegetables, and meat within the previous eight hours. However, he did have an arrow-point lodged in his back near one shoulder—a wound sufficiently serious to be fatal. He also had multiple cuts on his hands and arms, including one that penetrated to the bone. In addition, he had experienced a recent blow to the head. Clearly, there were other humans around at the time and place of his death.

Ötzi's remains are now interred in a climate-controlled laboratory in northern Italy, reflecting the fact that the site of his recovery was actually 304 feet (92.6 m) on the Italian side of the Austrian-Italian border. It also turned out (although this did not play into the legal jurisdiction argument) that Ötzi's last meal indeed came from the Italian side of the mountains.

GLACIAL MELTDOWN

The 1991 discovery of "Ötzi the Iceman" triggered a flurry of scientific as well as public interest. One question kept coming up: Why did Ötzi emerge from the ice *now*, of all times?

To try to answer this, there is something that we need to understand at the outset: Everything a glacier swallows up will eventually be regurgitated—whether it be a meteorite, a woolly mammoth, a beer can, or, yes, an Ötzi.

Glaciers are rivers of ice. They flow downhill in accordance with the laws of gravity and friction, albeit very slowly—usually mere inches per year. At their upper elevations, they accumulate new snowfall, which eventually compacts into ice, and this weight pushes them forward. If the underlying rock fails to provide enough friction to hold this weight back, we get an avalanche rather than a glacier. When, however, there is an approximate balance between the force driving the ice downhill and the friction retarding that motion, we get a glacier.

Glaciers never actually retreat in the sense of moving backward. If a glacier shrinks, it's only because its lower reaches are thawing faster, on average, than its upper levels are being replenished with snowfall. The actual movement of the ice, however, is always downhill. So even a perfectly healthy glacier will always eventually—over many thousands of years–expel everything it's consumed in the past. That natural expulsion will take place at the glacier's bottom fringe, or terminus.

Ötzi emerged facedown in a small pool of glacial meltwater at an elevation far, far above the glacial terminus. In other words, he surfaced tens of thousands of years before he should have. Clearly, the glacier that had encased him was thinning.

Was this some sort of local anomaly? Or are other glaciers also thinning? Various artist renderings of Alpine glaciers date back several centuries, and they indeed suggest that many glaciers were much broader and deeper in the past. Yet we cannot be sure that those long-dead artists were faithfully depicting reality. Photography, however, is a more verifiable medium (or at least it was, in the days before digital image manipulation). Photographs of many European glaciers date as far back as the 1850s. In virtually all of those instances, it is clear that those glaciers have retreated considerably—sometimes by as much as several miles—since those old photos were taken.

When a glacier is dynamically stable or expanding (rather than retreating), it behaves like a bulldozer, pushing rocky detritus ahead of it. When a glacier retreats, that line of rubble—called the terminal moraine—remains behind as telltale evidence of its maximum progress. As for thinning, geologists can usually also deduce that phenomenon by examining patterns of scouring on the walls of glacial ravines. In fact, this kind of fieldwork was begun, piecemeal, decades before Ötzi emerged in 1991. Since then, improvements in satellite telemetry have made the task much easier. Such research reveals that today, *most* of Europe's glaciers are simultaneously retreating *and* thinning.

According to the World Glacier Monitoring Service, 103 of the 110 glaciers in Switzerland, 95 of the 99 in Austria, all of the 69 in Italy, and all six in France are melting faster than they are being replenished by new snow accumulation. Moreover, it turns out that this deglaciation is not confined to Europe. In Asia, 67 percent of the thousands of Himalayan glaciers are retreating, as are about 95 percent of those in China. Today, in Glacier National Park in the United Sates, glacial ice now covers less than 25 percent of the area it blanketed back in 1850; within the next two decades it may disappear completely. In Alaska, a full 99 percent of the thousands of glaciers are shrinking. In fact, in Juneau, the glaciers have retreated so dramatically that the underlying soil, unburdened by the weight of the ice, has rebounded upward and actually created new expanses of coastal land beyond the reach of high tides. In equatorial Africa, the glaciers of Mount Kilimanjaro have shrunk by about 80 percent over the past century; at the current rate of melting, they could disappear completely by 2020. This pattern of glacial shrinkage even extends into the southern hemisphere, in New Zealand and southern South America. Glacial melting, in other words, seems to be a worldwide phenomenon.

1941

2004

**FIGURE 1: MUIR GLACIER IN GLACIER BAY
NATIONAL MONUMENT, ALASKA**
Glaciers all over the world are melting, but particularly in the northern hemisphere.

DISAPPEARING SNOW COVER

But it is not just the glaciers that are melting. In 2003, the western regions of Europe experienced an unusually warm summer. Before that date, most of the surface soil at elevations above 9,500 feet (2,896 m) was not only frozen solid, but it had been that way continuously since the last ice age. On steep slopes, it is this permafrost that stabilizes the dirt and anchors otherwise loose rocks. In 2003, however, the melt-zone climbed as high as 15,000 feet (4,572 m). The result was a tragic summer of landslides, rockslides, and avalanches that killed at least 50 climbers and injured or trapped hundreds of others who needed to be rescued. Since that blistering summer, the alpine permafrost line has recovered only modestly. It has not returned to its earlier lower levels, nor is it likely to do so in the foreseeable future.

Given that most of us are not mountain climbers, does it really matter that Earth's mountains are thawing? It turns out that, yes, it matters a great deal to a great many people. We will return to this subject later in the book, but it needs to be explained here that warm mountains do not hold snow very well. Yet thousands of cities and towns around the globe are dependent on meltwater from mountain snowcaps for hydroelectric power and/or drinking and irrigation water. In an overall warmer world, winter precipitation is no longer being partially stored on mountaintops—rather, it runs off almost immediately, often causing downstream flooding. Then in the summer, when the same streams would normally be sustained by meltwater from the snowcaps, they run low or even go dry.

If those same watercourses are also used to generate hydro-power, there may be seasonal electrical shortages or outages. If those streams are used for irrigation, there may not be enough water to sustain the crops through the dry season. Finally, if humans and/or wildlife depend on those disappearing streams for drinking water, the prognosis is obvious and ominous.

This is not just conjecture. It is already starting to happen in numerous places around the world. In the United States, seasonal water shortages are increasingly common on the west coast. Britain, France, Italy, Spain, and Portugal have also all had water shortages in recent years. Australia has suffered droughts affecting agriculture and stock keeping. In India and China, the seasonal scarcity of water is beginning to threaten the health of millions—and this relates only to regions dependent on melt-water from snowcaps, not to other causes of drought.

DWINDLING ICE CAPS

As important as they may be to downstream communities, mountain snowcaps account for only a small fraction of our planet's total ice cover. About 10 percent of Earth's land surface is buried beneath two huge, continental-sized ice sheets, thousands of feet thick—one blanketing most of Antarctica, the other covering most of Greenland. Combined with floating sea ice, these great ice sheets presently hold about 68 percent of our planet's fresh water (or about 1.7 percent of its total water). Alarmingly, large parts of both of these great continental ice sheets are melting, too.

Then, at the very top of the world, we have the Arctic ice cap. Figure 2 shows a series of top-view maps of this floating mass of ice from September 1980 to September 2008. During this 28-year period, this "permanent" ice cover shrank from 3.01 million square miles to 1.81 million square miles—a decline of nearly 40 percent. Although the rate of shrinkage is not consistent from year to year, the overall trend is abundantly clear. By 2030, the Arctic Ocean may be essentially ice-free during the summer.

The facts are beyond dispute: Earth's ice is disappearing—and whilst this may not be everywhere, it is definitely nearly everywhere. Overall, the sum total of the ice that melts each year vastly exceeds the amount of new ice that forms. So where does the difference go? Obviously it is still around, but it has become liquid water, which has physical properties quite different from those of ice. In other words, the shrinking ice cover is altering the physical dynamics of the entire planet. This is not conjecture, either; it is irrefutable, physical fact.

To begin with the most obvious difference, water flows quickly, whereas glacial ice barely creeps. For another, liquid water evaporates more rapidly than ice. (Yes, ice does evaporate; the scientific term is that it "sublimes," evaporating without first melting and becoming water; but it does this very slowly.) Ice and water also conduct and store heat differently; it is easier to change the temperature of a block of ice than an equal mass of water. And then there are differences in optical properties: Ice is generally more reflective than water. Scientifically, this reflective property is called the albedo.

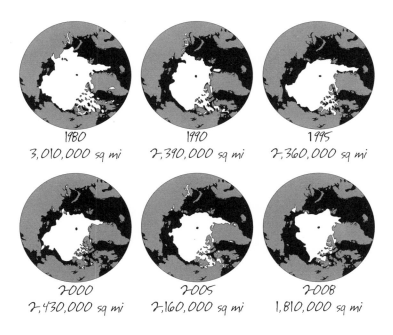

1980
3,010,000 sq mi

1990
2,390,000 sq mi

1995
2,360,000 sq mi

2000
2,430,000 sq mi

2005
2,160,000 sq mi

2008
1,810,000 sq mi

FIGURE 2: SHRINKING OF ARCTIC SEA ICE
Arctic sea ice from 1980 through 2008, as viewed from a point directly above the
North Pole. North America is at the bottom of the maps. The polar ice cap has shrunk
by nearly 40 percent in the same period and may disappear completely by 2030.
(Data from NOAA.)

For ocean ice, the albedo is 50 to 70 percent, and that of fresh snow is 80 to 90 percent. In other words, these percentages represent the proportions of incident sunlight that they reflect. Seawater, on the other hand, reflects relatively little sunlight, except near the times of sunrise or sunset. When the Sun is more than 30 degrees above the horizon, the albedo of seawater is less than 7 percent. The other 93 percent or more of the solar energy striking the seas is then absorbed—and converted to heat.

Furthermore, there is the matter of the phase transition. If you have ever thawed a frozen turkey, you know that the process is painfully slow. The reverse process—freezing—is also slow, as you can easily verify by putting a tray of water in your own freezer and checking on it from time to time. It turns out that the processes we call freezing and thawing both involve a rather large energy transfer. While it takes only one British thermal unit (1 BTU) to heat one pound of liquid water by a single Fahrenheit degree, melting one pound of ice requires 144 BTUs. The same applies in reverse: To freeze one pound of water, even after it has already been cooled to 32°F (0°C), you need to withdraw 144 BTUs of heat. This much heat simply cannot be transferred instantaneously, which is why thawing and freezing are such slow, sluggish processes. So, as a result, there is always a lag time between warming and melting, as well as between cooling and freezing. This is why it would be impossible, for instance, for us to wake up to a morning news broadcast informing us that the polar ice cap had suddenly disappeared overnight. If the ice cap indeed melts away, the process will be considerably slower and the initial evidence fairly subtle.

When it comes to the global climate, these and related factors interact in complicated ways. With a reduction in ice cover, incoming solar energy is increasingly absorbed rather than reflected from Earth's surface. Because most of our planetary surface is made up of water, this results in it being mainly the seas rather than the land that grow warmer. But this also increases the rate of evaporation, which is a *cooling* process. Moreover, much of the resulting increase in atmospheric moisture condenses into clouds, which not only alter the weather, but also affect the planet's overall albedo in complicated ways (higher clouds have higher albedos). With so many changes taking place simultaneously, the long-term impact on Earth's climate is monstrously difficult to predict mathematically.

As we shall soon see, however, such predictions have indeed been made—mostly through the use of sophisticated computer models, but also by extrapolating real data trends into the future. The results are far from encouraging—and it is common for these projections to err on the *low* side. In reality, our global climate may actually be rushing headlong toward a "tipping point," which could irretrievably change our planet into a much different place.

PUZZLES AND PONDERINGS

Of course, we also need to be cautious about jumping to unwarranted conclusions. The recent thawing of glaciers, snowpacks, and ice sheets raises quite a few questions. Among them are the scientific issues set out below.

- Scientists have been paying close attention to glaciers and ice sheets for only about 60 years. Relatively speaking, this is not very long, considering that the last great ice age ended about 11,000 years ago. How do we know that the present thawing is not just a short-term fluctuation that will soon correct itself?

- Why is the ice melting? Is the whole planet getting warmer, or are just certain spots warming up? (For example, only places at high altitudes and extreme latitudes.)

- If the planet is indeed warming on average, why is it doing so? Further, if this trend is real, does it have anything to do with human activities? How can we be sure?

- What future climate changes may already be locked in, given the lag times between the physical causes and their ultimate effects on climate?

- Earth has dozens of different climatological regions, within which their respective residents (humans, plants, and wildlife) have already adapted their lifestyles. How will these regions be affected in the future? Which will get dryer, wetter, warmer, or colder? Which might experience more extremes of weather, e.g., tornadoes, hurricanes, blizzards, droughts, and so on? What are the human and social implications of such changes? What are the biological and environmental implications?

- What climate changes are possible but uncertain? Why are they uncertain? What additional research needs to be done to resolve the outstanding questions?

- Does our global climate indeed have an irreversible "tipping point"? If so, when might it be reached? And what would be the consequences?

- Can an engineering initiative stabilize Earth's future climate? If so, to what extent and with what potential downsides? What would such an effort cost? Who would pay? Who would benefit? How might the various stakeholders reach agreement on a plan of action?

- What will be the likely outcome of doing nothing other than business as usual?

THE HUMAN CHALLENGE

So, you the reader are now invited to follow me in an exploration of the mechanics and prospects of global climate change.

But first, one advance warning: This is an interdisciplinary odyssey. It involves history, archaeology, geology, physics, astronomy, chemistry, biology, statistics, and a smattering of psychology and sociology. It is also not a linear quest; sometimes one line of thinking surges ahead of the others. No single discipline, it turns out, can claim any monopoly on truth or error about the global climate.

Even if you are only marginally concerned about the future of our world's climate, please bear with me. Although it is impossible to omit politics completely from the following pages, I promise to limit political issues mainly to the book's final chapter, where I discuss the current international treaties and protocols in place at the time of writing. However, political decisions cannot (or, at least, should not) be made either in a

vacuum, or in a quagmire of misinformation. As a global society, our basic challenge is to make sense of the complete bulk of the current scientific evidence, then—which is sometimes even more challenging—respect Mother Nature's whispered messages about the future of our planetary home.

CHAPTER 2

How Do We Know
What We Think We Know?

Human history becomes more and more a race between education and catastrophe.
H.G. WELLS, *The Outline of History*

There was a time, only a handful of centuries ago, when our Earth was considered to be the center of the universe. Aristotle had said so, the Hebrew Bible said so, and a long string of Catholic popes confirmed it; so it was true. The Creator had placed mankind at the hub of the cosmos, and that was that.

STARGAZERS

In the early 1500s, however, a Polish cleric named Mikolaj Kopernik began to muse about the apparent contradictions between the conventional wisdom and the observed motions of the planets in the night sky. If our Earth was at the center, then why did Venus never stray more than 47 degrees from the Sun? Why was Mercury always found within about 22 degrees of the Sun? And why do the other three visible planets—Mars, Jupiter, and Saturn—traverse the entire night sky, yet reverse their movements every now and then?

Kopernik played around with a series of diagrams. One of his sketches triggered an intellectual lightning bolt. When he set the Sun at the center, then drew six properly sized circles around it and dropped Earth in the third circular orbit, the whole scheme suddenly made sense. No longer did any planet actually reverse its motion; they all moved quite regularly in circles. To observers on the moving Earth, the other planets would only *appear* to be reversing from time to time. Writing in Latin under the pseudonym "Copernicus," the cleric published his findings in the year of his death, 1543. With that, our Earth was irreversibly demoted from the center of everything to simply the third rock from the Sun.

Since then, astronomers have further marginalized our planet's status. They now assure us that Earth is a mere speck that orbits an obscure star near the outer edge of a swirling cluster of millions of stars—the Milky Way galaxy—which, in turn, is an undistinguished one of billions of such galaxies in the known universe.

So what does this have to do with global climate change? The clue lies in the word "global." The discoveries of Copernicus and of other astronomers notwithstanding, most of us still do not really think of our planet as a planet. After all, rarely do many of us even glance up at the night sky—and when we do, what greets our eyes is a swarm of tiny points of light, not a vision of the awesome cosmos of which our own Earth occupies but an infinitesimal spot.

Anthropocentrism—putting humans at the center of everything—is very much our hallmark as a species. In fact, this prejudice has served us well in our survival on the planet (at least, more often than not) over the past 50,000 years or so. Only in the last few decades has such thinking begun to serve us badly.

Yet if our status in the cosmos strains our attention span, do we do any better when we think about the continuum of time? Not really, would seem to be the answer. In fact, our individual memories extend back only to our personal childhoods. On average, by the time we die, that has been a typical span of maybe 75 years. Nothing that happened prior is knowable firsthand. If there existed something before us, we take that on faith—faith in our elders, our history books, our religions, our myths, and maybe our science. So, while most of us assume that our surroundings will continue to exist after we die, we also take that supposition on faith, not on any firsthand knowledge of the future. We are not only stuck in our *here*, but we are also mired in our sense of *now*. Thus we consider it to be mere common sense to extrapolate our planetary present into the future, as if the essentials are incapable of changing. Our planet is going to be forever—right?

THE NATURE OF KNOWLEDGE

The topic of global climate change, however, transcends both geographical space and human lifespans. This is a subject where the notion of "common sense" often becomes nonsensical. Modern astronomy (including space exploration and planetary science) offers much to guide us, as do a variety of other specialized scientific disciplines (paleoclimatology, for instance). However, no single human has all the answers as to what is about to happen. Indeed, it is doubtful that anyone even has all the right questions.

What we do know, however, is that anthropocentric (human-centered), temporal, and "common-sense" thinking are woefully deficient tools with which to predict the future of the global climate—let alone what might be done to avert anthropogenic (human-generated) climate catastrophe, should such a task even be possible.

To begin, we need to think about what constitutes relevant evidence. How do we know what we think we know about climate? How do we know about the climate of the remote past, way before we were born, before people knew how to write (let alone measure temperature), or even before there were any people? Furthermore, on what basis do we think we can predict, or even rationally speculate about, the future climate of our planet—the climate after we personally have died?

These are not simple questions to answer. The whole subject requires a considerable amount of big-picture thinking—thinking that transcends our own small boxes of personal space and our own limited lifespans.

THE HOME PLANET

Our Earth is 7,926 miles (12,756 km) in diameter and about 25,000 miles (40,000 km) in equatorial circumference, depending on how you deal with its bumps and bulges. One way to gain some insight into the reality is to imagine our whole planet shrunken to the size of a basketball (with a 30-inch [76-cm] circumference) or—which works out pretty much the same—as a soccer ball (27–28 inches [69–71 cm] in circumference). Now add Earth's oceans and atmosphere to the ball, but do so in proportion to its shrunken size. How deep is our ball's ocean? How deep is its atmosphere?

The answers may come as a surprise. The oceans on the ball would average only 0.017 inches (0.4 mm) in depth, equivalent to the thickness of about four pages of this book. Those oceans would barely wet our palms. Even the deepest point in the sea (the Mariana Trench in the Pacific, at 36,198 feet [11,033 m]) would be the thickness of just 13 of these pages placed on such a ball. You might wet a fingertip there, but that is about all.

As for our ball's atmosphere, that is a trickier question. Because Earth's atmosphere becomes increasingly thinner with altitude, it is a matter of some judgment where the atmosphere ends and space itself begins. But half of our atmosphere does lie beneath an altitude of about 18,000 feet [5,486 m] which on our ball becomes a layer just seven pages thick. To be above 99 percent of the atmosphere (an altitude of about 20 miles [32 km]), we'd need to be only about 0.125 inches (3 mm) above the surface of a basketball-sized model.

Sally Ride, who in 1983 became the first woman in space, remembers the first time she peered out of her capsule window for a view of her home planet. Although totally familiar with the arithmetic relating to the scale of the atmosphere, she was nevertheless awestruck by the sight of that pencil-thin blue crescent clinging to the globe's curvature. Earth's atmosphere and its oceans are not nearly as vast as our "common sense" may suggest. In fact, they are quite thin and delicate. Yet all of our planet's weather activity and climate is imbedded in these two fragile blankets; and furthermore, they are mostly in a small portion of them—the lower half of the atmosphere and the upper 100 feet (30 m) or so of the seas.

How Do We Know?

A few paragraphs ago, I quoted data relating to the size of Earth, the average depth of the oceans, and so forth, along with some straightforward arithmetic to scale everything down to the size of a basketball. But how do we know that the original numbers are accurate? After all, which of us has ever personally measured the depth of the oceans or the height of the atmosphere? Lacking firsthand knowledge of these figures, how can you have any confidence in anything you are reading here?

This is far from a frivolous issue. In fact, it strikes at the heart of the decades-long controversies about global climate change. How do we know what we think we know? How well do we know those things?

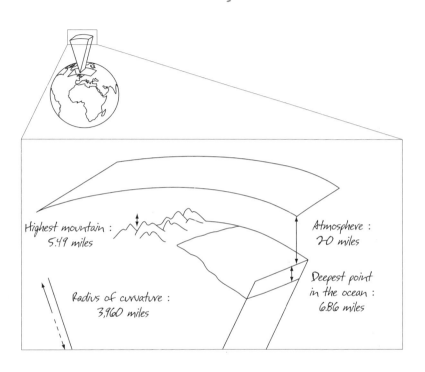

FIGURE 3: THE ATMOSPHERE OF EARTH
On the scale of the planet, Earth's atmosphere is an extremely thin blanket—and its composition is being changed by human activity.

These are important issues to which we will need to return from time to time in the following pages. To begin with, let us consider the six "ways of knowing" identified by the psychologist G. C. Helmstadter back in 1970. This is not a perfect system, and there are some overlaps; but on the whole, his scheme works pretty well. Let us briefly examine each of these six modes.

TENACITY

To Helmstadter, tenacity is the acquisition of knowledge—sometimes involuntarily—through repetitive delivery. Television advertisers, for instance, routinely run one commercial over and over rather than changing it, because such repetition sinks their product name or jingle deep into the viewers' minds. Some of the less scrupulous political leaders throughout the world apply the same principle: Repeat a falsehood often enough and eventually it becomes indistinguishable from truth. Tenacity, however, is not always invalid. For instance, it is the way teachers learn their students' names and the way all of us learn new languages. Knowledge gained this way, however, seldom involves critical thinking. After all, this is basically the way parrots learn to mimic. Most of the early propaganda against the science of global warming and its conclusions exploited the principle of tenacity: Repeat, repeat, and do not critically examine any new evidence that is emerging.

INTUITION

This "way of knowing" is connected to our inner feelings. At its best, intuition can aid our survival. For example, without even thinking about it, we probably slow down when driving in a rainstorm, or we intuitively avoid a rustling bush in the forest. Mysticism and spiritualism also fall into this category, leading to a very personal kind of knowledge, which may or may not be true for others. Regardless of their validity, there is little you can do to transfer your intuitions to anyone else, unless the other person is already prone to feel the same way—in which case, both will agree those mutual intuitions are simply common sense. For individuals who feel fervently about an ideology, be it religious or political, intuition can be a very strong driving force and can blind the mind to arriving at more objective truths. Many people seem to have rejected the early evidence of global warming on the basis of ideologically driven intuition.

AUTHORITY

Authority is the acceptance of an idea because a respected person or group asserts that it is true. Most of the formal learning that takes place around the globe falls into this category. Textbooks, Internet sites, TV news shows, magazines, instruction manuals, religious leaders, and so on all purport to be authoritative sources. Yet nobody's assertions are ever 100 percent accurate. Indeed, some can be seriously inaccurate—through human error, unexamined assumptions, or sometimes even through overt dishonesty. Yes, that anonymous Internet blogger could possibly be right when all the Nobel-winning scientists are wrong; but how can we decide?

RATIONALISM

Rationalism is the creation of new knowledge through logical reasoning based on existing knowledge. Civil engineers, for instance, are remarkably successful at predicting whether a new bridge, even one with unique design elements, will ultimately function as planned with no bad surprises. Most police investigations and legal systems are also set up to function rationally (although whether they always succeed is another matter). Probably the purest example of rationalism is Euclid's system of geometry, which starts out with a small set of axioms and postulates. It then derives logically thousands of conclusions about various geometrical relationships, many of them rather obscure. Rationalism, properly applied, has demonstrated itself to be an extremely credible method for generating new knowledge from old knowledge. Although they may originate in a hodgepodge of questions and answers, scientific theories are almost always recast into a rationalistic format before they are disseminated to other scientists for review and evaluation.

EMPIRICISM

Empiricism is the process of learning through experience and observation. In its simplest form, a child may discover empirically why you should never put a raw egg in a microwave oven. Or a college student may learn (as one once explained to me) why you should never punch a fat guy in the stomach. (If you do not know what I mean by this example, it is because you have never experienced something like that yourself—which is exactly my point.) Yet our natural human senses are limited in scope and sensitivity, and this has led numerous curious individuals to invent instruments such as telescopes, microscopes, voltmeters, spectrometers, seismographs, radiation detectors, and so on. Alas, such sensory extensions may sometimes encourage errors (as was the case when in 1895 Percival Lowell reported seeing through his telescope what he took to be canal-like structures on Mars). Still, there is generally a level of truth in seeing—and yes, videos are one example of an instrument for enhanced seeing.

SCIENCE

This is the process of probing the mind of Mother Nature in order to tease out her secrets. Science is not simply the names of the chemical elements nor the vocabulary for the parts of the human body; terminology like that is known through authority, not science. Science is the *inquiry* into why, when, and how the universe behaves as it does. And, as such, science is a synthesis of rationalism and empiricism that involves observing, analyzing, theorizing, predicting, speculating "what if," and, most importantly, rejecting erroneous ideas.

MERGING WHAT WE KNOW

Obviously, the six modes of knowing just explained are not mutually exclusive; human knowledge is often based on some combination of them.

The Wright brothers, for instance, seem to have known *intuitively* that it was possible for us to fly; they then checked the writings of past *authorities* on the subject, some of which were wrong; they studied the flights of birds *empirically*; they *rationally* analyzed the factors that explain avian flight; they built a wind tunnel and conducted *empirical* investigations both in the lab and the field; and they rejected bad ideas (of which they had more than a few of their own). Ultimately, however, it was not the Wrights' intuition, deference to authority, or even their empiricism and rationalism that clinched their success. It was their science—their firsthand inquiries into the why and how of powered heavier-than-air flight.

The Wrights' stunning achievement of 1903 was not widely acknowledged by the public for several years. Most people placed more faith in their own tenacious knowledge to the contrary, fueled by the oft-repeated chestnut: "If God had meant men to fly, He would have given us wings." And yes, the impossibility of manned flight was consistent with everyone's intuition at the time. To compound matters, there were somber pronouncements by various authorities in the newspapers that the Wrights could not have possibly achieved what the first few eyewitnesses said they had. As for rationalism, learned scientists around the globe (including several at the Smithsonian Institution) applied the then-understood laws of aerodynamics and concluded that the Wright brothers' machine was incapable of controlled flight!

The history of global climate research has followed a remarkably similar pattern. Millions of people initially based their climate change conclusions on tenacity—they heard over and over from various sources that the planet's climate is *not* changing, never mind what the scientists said. For many members of the public, this conclusion was reinforced by intuition. After all, it did not *feel* like it was getting much hotter or rainier from year to year, therefore common sense said it was not. Some went a step further and sought authority. All authorities, however, do not have equal veracity, and sometimes it can be very hard to distinguish the competent ones from the incompetent (or even fraudulent). Many of the public concluded that the experts were issuing nothing more than intuitive opinions—opinions which ran the gamut from yes, Earth is warming catastrophically, to the claim that temperatures are relatively stable or actually dropping (which, in fact, they have—in some parts of Antarctica, for instance). Some naysayers went beyond authority, however, and argued—using rational logic—that the sea and the atmosphere are both too large to be affected by mere human activity, and that any perceived change in Earth's climate amounts to no more than normal statistical fluctuations.

Which brings us to empiricism. In the last several years, we have all seen videos of polar bears stranded unseasonably on melting ice floes, while in temperate climates growing seasons are getting longer. Warm-weather plants are creeping northward, and are being found at higher altitudes than ever before observed. Bird watchers are documenting environmental conflicts between migrations and the availability of avian food sources, which are both triggered by unseasonably early spring thaws. Other corroborating evidence of climate instability accumulates almost daily.

We are now at the point where we can discard critically all previous disclaimers about the reality of climate change. So, with what do we replace those discredited negatives? Well, hopefully, with scientific perspectives. Please note, however, that there are not detailed and definitive scientific answers to everything—because with every scientific conclusion, there comes one or more caveats and qualifications. What I am saying, though, is that science is the best approach we have for attempting to understand the intricacies and mysteries of our universe, our environment, and our climate.

REAL SCIENCE

But what exactly is science? One wag has written that it is what scientists do late at night. Some textbooks state authoritatively that it is what you get when you follow the five or six steps of something called the "scientific method" (with apologies and sympathy to the students who took the time to memorize that list, as there happens to be no such thing as *the* scientific method). In addition, there is the view that science is merely nomenclature, e.g., the classifications and labels on the rocks in the geology exhibit at the local museum, or the names of the bones and muscles of the human body.

Science is none of these. Certainly, it is not about being a designated human authority. Nor is it about following a set of rules, nor about vocabulary—and it is not focused on what everyone already knows. Science is actually about *ignorance*. Science deals with what we do not yet know: the unanswered questions in Mother Nature's chest of secrets.

To be a scientist, you need to be ignorant about some phenomenon. You also need to admit your lack of understanding, and to have the intellectual curiosity to reconfigure your intellectual vacuum as a puzzle that might be solved. Then you need to devise some creative way to get Mother Nature herself to cough up a plausible answer—perhaps through controlled laboratory experiments, maybe through field studies, or maybe through some combination.

But even when you do arrive at a plausible answer, you're far from done. Your next task is to replicate your findings, establishing that your results are not merely subjective but that they have a firm basis in objective reality. Therein lies the big challenge; for no scientist personally lives long enough to study the whole universe over all time.

Thus, ultimately, science necessarily becomes a social endeavor, particularly in interdisciplinary fields like the dynamics of the global climate. It requires the cooperation of many, many people conducting many, many studies, usually over a considerable period of time.

CHAPTER 3

Climates, Climates Everywhere

Climate is what we expect. Weather is what we get.
MARK TWAIN

When Alexander the Great embarked on his campaign of world conquest in 336 BCE, he and his soldiers were astonished by the diversity of geographies and human cultures they encountered. In particular, never before had such a large group—at least 42,000 men—experienced such an assortment of climates firsthand, ranging from arid deserts to steamy forests to the snowy slopes of the Himalayan foothills. Although Alexander himself did not live very long—nor did his empire—the fact that climate varied widely from place to place was clearly established. Over the following centuries, an ever-expanding collection of such observations was enshrined in the great library at Alexandria, where those scrolls remained accessible to scholars until the library's destruction.

Those early observations, of course, had to be subjective. The thermometer was not perfected until the 1500s, and even after that, the only temperatures that tended to be documented related to extreme events such as heat waves or unseasonable chills. It was not until after the invention of the telegraph and, more particularly, the laying of undersea cables in the latter part of the nineteenth century that more comprehensive weather data began to be collected and disseminated on a global scale. Only after such meteorological records had been accumulated for many decades was it possible to start thinking about climate in quantitative terms.

CLIMATIC REGIONS

Climate, as the term is usually used today, is the average of regional weather over at least a 30-year period. (However, in some instances, scientists may also talk about longer time spans— say, up to a century.) By "weather," we mean mainly air temperature and all kinds of precipitation, but we may also include wind data and, in the case of coastal cities, sea conditions. Although it would be tidy to standardize the meaning of "region" in this definition, our Earth tends not to be particularly cooperative on that score. For instance, there is a 5,000-mile (8,000-km) swathe spanning North Africa from the Atlantic coast into and including most of the Arabian peninsula that is nearly all arid desert with meager or no rainfall, and so it can be regarded as one single climatic region. Other climatic regions, however, can be quite small, like the 150-mile-wide (240 km) pocket of Mediterranean type dry-summer climate found at the Cape of Good Hope in South Africa. As a result, the climate map of the world is a patchwork of strangely shaped regions of wildly different sizes.

Such regional climate is sometimes referred to as "macroclimate" in order to distinguish it from mesoclimate, microclimate, and global climate.

Mesoclimate in particular is worth mentioning, because its variations can sometimes foster false conclusions about macroclimate and/or global climate. By "mesoclimate," we mean a relatively small region whose average weather deviates noticeably from the surrounding macroclimate. Typically this happens in cities, which may be warmer, windier, or possibly foggier than the surrounding countryside. As a result, weather stations installed in such places may not reveal a true picture of the climate of the rest of that region.

Even more specific is the use of "microclimate," which relates to even tinier spots. For instance, one may notice a plant flowering or a tree budding unseasonably, where the agent turns out to be a streetlight or maybe a natural spring that is keeping it warm. Conversely, a small patch of ground that never gets sunlight may freeze on an evening when its surroundings do not. Again, the issue here is that we not let such micro-anomalies seduce us into jumping to erroneous generalizations.

Which brings us to "global climate." How does one tackle such a concept, given that geographers identify at least 125 regions of macroclimate on the globe? Each of those distinct regions is contiguous with one or more other climatic regions, yet all those boundaries are presumably capable of shifting, just as the Sahara Desert and the bordering grasslands (the Sahel) are currently creeping southward into places that were once tropical rain forest. So why should we even create a fuss around the concept of global climate, when it is regional macroclimate that we actually experience?

The answer is that a global view can reveal climatic changes and trends that may not be apparent when we focus more narrowly on specific regions. Our Earth, after all, is a single planet, and everything that happens here is interconnected. If one fragment of the world's climate changes, there is bound to be a planetary ripple effect—maybe a minor one, but maybe one that is not so minor. The only way we can hope to tell is to look at the big picture.

A TALE OF THREE PLANETS

Our Earth has two companion planets, quite similar in some respects, but astoundingly different in their climates. Both are visible via the naked eye. One is Venus, the third-brightest object in the sky (after the Sun and Moon), with a diameter 95 percent that of Earth. The other is Mars, whose diameter is slightly more than half that of Earth's, and which, under favorable conditions, appears in the night sky as a reddish-orange object of medium brightness.

VENUS
Venus orbits the Sun at a distance of 72.3 percent of Earth's orbital radius. As a result, it receives 1.9 times as much incident

sunlight as does our own planet. However, Venus's constant and complete cloud cover (droplets of sulfuric acid floating in a mist of sulfur dioxide) reflects some 76 percent of this incident solar radiation right back into space, which is why the planet appears so bright. The amount of solar energy penetrating those sulfurous clouds and actually passing into Venus's atmosphere is only half as great as that which penetrates Earth. In fact, Venus's solid surface is rather dimly lit. If this dribbling solar input were the only factor affecting that planet's surface temperature, Venus would be a rather cold place—certainly chillier than Earth.

That, however, is not the case. Instrumentation on two Russian robotic landers, as well as spectroscopic data from various flyby missions, have consistently measured Venus's surface temperature as being within the range 860–896°F (460–480°C). This is true on the night side as well as the daylight side (even though night on Venus lasts 121 Earth days). Because Venus's entire atmosphere swirls around the planet once every 100 hours—much faster than the planet's period of rotation on its axis—its climate does not vary much from place to place. Nevertheless, Venus does seem to have weather, with lightning and even thunder. There have even been photographs taken there showing sand dunes and rocks weathered by wind erosion.

It is Venus's atmospheric composition that gives us a clue as to why its climate is so drastically different from that of Earth. Spectrographic instruments tell us that the air there is made up of 96.5 percent carbon dioxide (CO_2) and 3.5 percent nitrogen (N_2), and that it is packed 92 times as densely as our Earth's atmosphere. Because there is no vegetation and no liquid water, there is nothing to soak up any of this Venusian CO_2. Thus, it stays in the atmosphere, where it behaves like a heavy bedspread that traps heat near the ground. The overall result is that, although relatively little sunlight gets through to the Venusian surface, that small amount of solar energy becomes trapped there very effectively. And so the planet has become hot—very hot indeed.

Was the climate on Venus always like this? It seems unlikely. Because Venus's crust is composed of materials that are very similar to those on Earth, and given its proximity to us, it seems likely that both planets were formed at about the same time cosmologically speaking, and through similar physical processes. It is quite plausible, therefore, that Earth and Venus started out with comparable atmospheres and maybe even similar amounts of water. Yet somewhere along the line, maybe three billion years ago or so, the two planetary climates began to diverge, with Earth developing an oxygen-rich but CO_2-poor atmosphere, and the Venusian atmosphere heading off in the other direction.

What would drive such a departure? Nobody knows for sure. Maybe it was simply a statistical difference in the relative abundances of various elements when the two planets initially condensed out of the primordial clouds of gases. Or maybe it was some other agent—say, later collisions with other interplanetary objects. Or then again, maybe it was life. The most distinguishing feature of planet Earth is not its chemical or physical makeup, but rather the fact that you cannot go anywhere on the globe without finding life—lots of life. Earth's biosphere is made up of forms of life whose ancestors not only arose quite early in the planet's prehistory, but which also from those earliest times were probably a driving force in creating the atmosphere of today. In other words, an atmosphere that contains enough oxygen to support animals as well as plants. And one that has relatively little CO_2, because so much of it has been sequestered in the cells of our planet's abundant flora.

However, I mentioned two sister planets. Let us look briefly at the second.

MARS

Mars orbits the Sun at 1.52 times Earth's orbital radius. As a result, the sunlight striking Mars is only about 43 percent as intense as on Earth. This may sound like a huge reduction in sunlight, yet it is roughly equivalent to comparing a summer day in Singapore to a winter day in Paris. Yes, Paris will be chillier, but not unbearable. Mars, however, is really, really cold. Temperatures there, as measured around five feet (1.5 m) above the Martian surface by the Viking landers, range typically from 1°F to -178°F (-17°C to -107°C). Furthermore, these were recorded in the planet's temperate zones.

So what's going on here? Why is Venus so hot, Mars so cold, and our own Earth so temperate?

Mars's atmosphere, it turns out, is less than one percent as dense as Earth's. Although it is composed of 95 percent CO_2 and 3 percent nitrogen (quite similar to Venus), there just is not all that much of it. As for water, yes, it is certainly there. In fact, there seems to be enough of it in the frozen Martian polar caps that if it were all to melt at once, it would flood the entire planet to an average depth of about 30 feet (9 m). Further, Mars has weather, too, and although not plentiful, also has clouds of frozen water and CO_2—not to mention its dust devils and wind gusts of up to 300 mph (500 kph). The planet is frigid mainly because its atmospheric blanket is so thin.

DOWN TO EARTH

A great deal remains to be discovered about the climates of Mars, Venus, and—yes—Earth. But what we have discovered already gives us plenty of food for thought. We know that these three worlds are similar in many respects, yet their climates are wildly different, and two of the three seem to be uninhabitable. On one, we find a heavy blanket of CO_2 and a surface hot enough to melt lead. On another, we find a sparse atmosphere and surface temperatures dropping low enough to freeze CO_2. On the third planet—our own—we have a patchwork of climatic regions ranging from cool to warm and wet to dry, most of which swarm with life. But will Earth always be so hospitable to living creatures?

Suppose that our trio of rocky planets indeed started out pretty much the same. Beyond the issue of why their climates differ so drastically today comes the deeper question of whether Earth's climate is truly stable over the long term. Are there factors that might tip it in one direction or the other, say, more toward the frigid conditions on Mars, or more toward the scorching conditions on Venus?

MATHEMATICS AND SCIENCE

The physical sciences do not create knowledge in the same way that mathematics does. For instance, using mathematics, we can prove that the angles of any flat (plane) triangle add up to 180 degrees. This is always the case; there are no exceptions.

A mathematician does not arrive at such a result by actually drawing a bunch of triangles and measuring them—because no measurement can ever be exact, and also because it is physically impossible to construct every imaginable triangle during one human lifetime. Instead, mathematicians prove their conclusions by reasoning logically from a small set of a priori axioms and postulates. The result is an airtight logical sequence that leaves no room for doubt about the conclusion—given, of course, that we accept the starting premises. Mathematics, in other words, leads to demonstrable truths. (It is worth noting that such truths are also always conditional. Draw a triangle on something other than a perfectly flat surface, for instance, and its angles will *not* add up to 180 degrees.)

As for scientists, unlike mathematicians, they can do no better than to arrive at plausible truths. There is no set of a priori axioms and postulates for our physical universe. And certainly there is no way any scientist can remedy the matter by observing the entire universe for all time. The best any researcher can actually do is to explore a few things here and there, now and then, and then attempt to construct the bigger picture. Sometimes the result is horribly wrong (as with the announcement of the discovery of "cold nuclear fusion" in 1989). More often, however,

the outcomes do contain kernels of truth—which in turn cry out
for further investigation.

AGASSIZ, FISH, AND AGE OF THE EARTH

The idea that Earth's climate has changed significantly over time
originated with the ideas of Louis Agassiz around 1837. Agassiz,
born in Switzerland in 1807, was fascinated by fish, which led
him to study not just living fish, but fossil fish as well. Variations
in the fossil skeletons he recovered from shallow and deeper
sediments convinced him that the watery environments of his
specimens had varied over time. Moreover, he noticed that
these evolutionary changes were dissimilar in different parts
of the world. In particular, it occurred to Agassiz that Switzerland's summertime lakes were brimming with meltwater from
glaciers, whereas Brazil's waterways were not. This sidetracked
Agassiz's curiosity to the then poorly-understood phenomenon
of glaciation.

Although several other naturalists had already speculated
that the erratic patterns of rocks strewn over the slopes of the
Jura Mountains had been moved there by past glaciers, Agassiz conducted much more extensive studies of the evidence—
examining patterns of striations and gravel moraines created by
numerous existing glaciers, then looking for similar patterns in
places where there were no current glaciers. His conclusion was
that one vast sheet of ice had once blanketed most of Europe.
There had been, in the distant past, a "Great Ice Age."

At that time, in 1837, nobody had a clear idea of the age

of the Earth. Many believed the scriptural analysis by the Irish Archbishop Usher, who in 1654 established the Year of the Creation as 4004 BCE. Many naturalists who marveled at mountains and studied features carved by the slow processes of erosion felt intuitively that our planet is much, much older; but as for coming up with a number, they were at a loss. Some creative thinkers even carried out the equivalent of an intellectual end-run and tried to answer the question by computing the age of the Sun.

Their answers: several thousand years at most.

Their reasoning typically went something like this: Numerous scientists had already measured the intensity of sunlight at the surface of our Earth (using calorimetry; i.e. timing how long it takes a measured amount of water to experience a measured temperature rise). The result on sunny days—in today's units—was always around 1,000 watts per square meter (W/m^2), more or less. Newton's law of gravity (1687) and his principles of orbital mechanics had already clearly established that the mass of the Sun is equivalent to about 300,000 Earths. The distance to the Sun was also known to reasonable accuracy ever since Giovanni Cassini's laborious planetary-parallax computations of 1672: about 93 million miles (150 million kilometers).

Now comes the intellectual leap: What is fueling the Sun? At that time, of course, nobody knew about thermonuclear fusion, so the only possible answer was chemical combustion—that the Sun is essentially a big bonfire in the sky. And a great deal was known about heats of combustion; this was, after all, during the Industrial Revolution when most factories in the Western world were being driven by steam. So—assuming that the Sun is a giant lump of burning coal—how long would that lump burn at the rate it is observed to be emitting heat? Answer: 2,455 years, more or less.

The Earth is clearly older than that. The Sun, therefore, cannot be burning coal; it must be burning something else. Yet no matter what assumptions were made about the fuel source (again, nobody knew about thermonuclear fusion then), the answer could never be stretched even to Archbishop Usher's modest figure for the age of the Earth.

So when did that "Great Ice Age" occur, if the planet was so young? Agassiz had no good answer. He returned to his interests in zoology, moved to the United States, married an American college instructor (Elizabeth Cabot Cary), taught at Cornell and Harvard, and by all accounts was loved and respected by everyone who knew him. He died in 1873 at the age of 66.

Shortly after Stanford University opened in 1891, a full-sized bronze statue of Louis Agassiz was installed on a pedestal above the elegant entrance to the zoology building. In 1906, the great San Francisco earthquake heaved that sculpture to the pavement, driving it headfirst through the concrete to its chest. On viewing the ignominious result, one of the Stanford faculty is alleged to have remarked, "Dr. Agassiz was great in the abstract, but not in the concrete."

DATING THE EARTH

Agassiz's ice-age hypothesis inspired numerous other naturalists to try to fill in the blanks. The more they investigated, the more information they came up with. To the disappointment of some, Agassiz's great ice sheet turned out to be somewhat smaller in extent than he had surmised. On the other hand, numerous field studies began to discover plant fragments interlayered with glacial deposits, and fossils of warm-weather animals like reptiles even in cold climes. This was the compelling evidence that there had been not just one "Great Ice Age," but multiple ice ages, punctuated by warm periods. In other words, Earth's climate had apparently fluctuated wildly at numerous times in the distant past.

This finding exacerbated that earlier, still-nagging issue of time. If indeed our Earth were only a few thousand years old, as both Holy Scripture and the thermodynamic calculations suggested, then why were there no historical records of all those nasty climate changes? Numerous weather anomalies were recorded by the ancients (droughts, floods, and so on), as were various astronomical events. But nowhere in recorded history was there any mention of the southerly crawl of great sheets of ice, plowing boulders and forests ahead of them. If something like that had happened during historical times, would it not have been noteworthy enough for someone to inscribe the event on a stone or a clay tablet—or at least to incorporate it into cultural myths and legends?

When discrepancies like this arise, something needs to give. Most scientists chose to reject the literal interpretation of the Biblical creation chronology, as well as the thermodynamically computed age of the Sun, in favor of other ways of arriving at the age of planet Earth.

In 1862, the British scientist William Thomson, Lord Kelvin, carried out a series of calculations based on the assumptions that our planet was originally molten, and that it has been inexorably cooling off through the processes of thermal conduction and convection. His results indicated that it had to have taken 24 to 400 million years for Earth's surface to cool to its present temperature.

Years later, John Joly's Dublin University–based investigations of the delivery of salt to the ocean via rivers led him to conclude that our seas must be 90 to 100 million years old. Meanwhile, others compared rates of sedimentation to the thicknesses of sedimentary rock samples and arrived at figures as high as 1.6 billion years for the ages of those stones. In 1896, Henri Becquerel discovered radioactivity. As more and more laboratory data were tabulated about the natural rates of radioactive decay of elements like uranium, thorium, radium, and so on, another dating strategy suggested itself. This was the idea of analyzing the chemical compositions of various rocks and calculating the ratio of the radioactive elements to their decay products, then working backward to compute how long this process had been going on.

The first time this was done, in 1905, Ernest Rutherford (at Manchester University in England) and Bertram Boltwood (at Yale) dated the origins of several rocks to about 500 million years ago. Two years later, using a more refined technique, Boltwood dated a rock sample to 1.64 billion years. That analytical process, however, was cumbersome; it involved traditional "wet" chemistry, tediously performed with no instrumentation other than a laboratory balance, and it left the nagging issue of whether a particular chunk of rock truly reflected the age of the whole planet.

After about 1918, however, the Rutherford-Boltwood technique was streamlined by the invention of the mass spectrometer, which can separate out the isotopic constituents of very small samples according to their atomic masses. As a long stream of additional rock samples was analyzed over the following 90 years, older and older ones showed up in the mix.

So far, the oldest seems to be a collection of small crystals of zircon found in western Australia. Using a technique based on the rates at which radioactive potassium-40 is known to decay into stable argon and calcium, the crystals' origin has been dated to 4.404 billion years ago. The samples probably originated very close to the time that Earth first developed a solid crust. The age of our planet seems to be only slightly older, with the currently accepted figure at around 4.54 billion years.

Although no single measurement is ever definitive, the plausibility of the truth of those results grows rapidly when a pattern of consistency arises from multiple measurements by many independent observers. This is the case with the age of our planet. It is probably not *exactly* 4.54 billion years or 4.404 billion, but it most definitely lies somewhere in or near that range. Thus, today, any claim that our planet is only thousands or even millions of years old is no more credible than a claim that the Earth is flat.

Clearly, then, there has been plenty of time for Earth's climate to change. But did it really? And if so, will it do so again in the future? And if that indeed happens, will it be good or bad for the future of Earth's biosphere?

CHAPTER 4

Changes of the Ages

The only thing constant is change.
ATTRIBUTED TO HERACLITUS OF EPHESUS

In 1858, the 50-year-old Charles Darwin received a letter from an unknown 35-year-old naturalist named Alfred Wallace. Wallace was fascinated by the question of why some plants and animals go extinct even as others evolve into new forms, and he wanted to share his hypothesis on the matter. Unbeknownst to the writer, however, Darwin had already been struggling with this same question for about 25 years and had developed a scientific explanation of his own—which was similar to Wallace's, but more refined. Concerned that he might not be credited with his own theory, Darwin quickly condensed and edited the manuscript that he had been preparing for the previous two decades. The following year, 1859, Darwin's *On the Origin of Species* first appeared in print.

BIOLOGICAL EVOLUTION

In the views of both Darwin and Wallace, the main driving force behind speciation is environmental change. During an ice age, for instance, vegetation growth shifts toward more temperate latitudes. This, in turn, alters the food sources available to various fauna and puts some of them at a survival disadvantage. Meanwhile, the expansion of continental ice lowers the sea level because ice that is caught up in glaciers is water that does not run off into the oceans. In turn, lower sea levels occasionally expose land bridges between previously isolated landmasses, allowing animals to migrate into new environmental niches, where some of them may well thrive better than ever before. Conversely, such access may also bring about a whole influx of new predators.

During warmer climatological periods, of course, the opposite set of events occurs. Rising seas isolate some life-forms on islands, while others on continents may gain new sources of nourishment. The species that survive over the long term are those whose biological machinery evolves in harmony with their changing environments. Species that cannot reproduce rapidly enough, or which lack the genetic diversity to change as rapidly as their environments, are relegated to extinction and the fossil records.

Some of the successful adaptations are quite remarkable, as both Darwin and Wallace pointed out. Examples include finches in the Galapagos Islands, where the birds' beaks are uniquely shaped to harvest particular types of fruits and nuts on specific islands; moths in England that are naturally camouflaged and

visually disappear when they land on tree trunks of a matching pattern; and bears that hibernate through northern winters without even urinating.

Of course, no individual animal actually modifies its own traits. Rather, it is the whole species whose characteristics shift, albeit gradually, over the course of many generations. Such shifts occur because there is always some degree of genetic diversity in every life-form, and the individuals who happen to be the best adapted are the ones that survive long enough to reproduce and pass on their advantageous features to the next generation. Darwin called this process "natural selection."

So why did Darwin wait so long to publish his theory? Because there was one nagging problem that he simply could not resolve to his own satisfaction: the age of Earth. At the time, in the 1850s, our planet did not seem to be old enough for the slow grind of natural selection to have had sufficient time to work its wonders.

However, as we have already seen in the previous chapter, the time-span problem would soon enough be resolved and Earth's solid crust seen to date back not the mere tens of thousands of years as was generally believed in Darwin's day, but a hundred thousand times as long.

So it turned out that the dilemma was resolved. Earth is more than old enough for Darwin and Wallace's sluggish creep of natural selection to have accomplished its wondrous deeds. By the early 1900s, the time scales for biological evolution began to mesh fairly well with the evidence for past global climate changes—and after all, it is environmental change that drives biological evolution.

Unsurprisingly, not everyone was convinced. Whilst some of the naysayers had actually read Darwin's and Wallace's writings, the majority of doubters had not gone to that degree of intellectual labor. Instead, they based their objections on hearsay, intuition, and/or religious conviction. (One should note here, however, that a number of Christian denominations—including Roman Catholics—took a hands-off approach from the very outset. The papacy had already been humiliated sufficiently by the sorry episode of the 1633 trial of Galileo for the heresy of claiming that Earth orbits the Sun. Church fathers were not interested in making another such embarrassing mistake, so they adopted the position that scientific research on matters of evolution, the age of Earth, and so on had nothing to do with religious faith or morals. The Catholic faithful were free to draw their own conclusions about such matters.)

By the end of the nineteenth century, the geologists lost their monopoly on the concept of global climate change. Climate issues began to attract the attention of many biologists plus quite a few chemists and physicists, and even many of those previously skeptical on a religious basis.

SOMETHING IN THE AIR

By the early 1900s, chemists had conducted numerous analyses of the composition of Earth's atmosphere. The principal component turned out to be molecular nitrogen (N_2), which accounts for about 78 percent of the atmosphere by volume. Next in abundance is molecular oxygen (O_2), at about 21 percent. The remaining one percent or so is composed of a mixture of trace elements. Currently accepted values are listed in Table I, expressed in parts-per-million (ppm)—the number of molecules per one million molecules of atmospheric air.

Table I: Permanent composition of dry
atmospheric air in parts per million (ppm)

Gas	Chemical Symbol	ppm
Nitrogen	N_2	780,800
Oxygen	O_2	209,500
Argon	Ar	9,300
Neon	Ne	18
Helium	He	5
Hydrogen	H_2	0.6

Table I, however, is not quite complete. In addition to the permanent gases listed, certain others are also found in variable concentrations. Two of them—water vapor and carbon dioxide— came as no surprise to the early chemists. Water vapor (H_2O) enters the atmosphere through natural evaporation. Carbon dioxide (CO_2) arises through combustion, the decay of vegetation, and through other natural processes like volcanoes and the process known as outgassing from the oceans—and even breathing.

Table II lists representative values for the most common of these variable gases. Of course there are also others: sulfur oxides near active volcanoes and some coal-fired power plants, for instance. In addition, the atmosphere contains particulates (dust, soot, and so on), which of course are not gases.

Table II: Variable atmospheric gases
in parts per million (ppm)

Gas	Chemical Symbol	ppm (typical)
Water vapor	H_2O	0–40,000
Carbon dioxide	CO_2	386 and rising
Methane	CH_4	1.7 and rising
Nitrous oxide	N_2O	0.3
Ozone	O_3	0.4 at the surface; 5–12 in the stratosphere

Normally, to the human eye, CO_2 seems just as transparent as nitrogen and oxygen. Yet what would this gas look like if our eyes could detect longer wavelengths—infrared radiation, beyond the red end of the visible spectrum? As Isaac Newton discovered in 1665, although we are unable to see anything in the infrared, we can *feel* it as being warm to the skin. Not only that, but if infrared radiation is directed onto a thermometer via a glass prism, the instrument will record a rise in temperature.

It is a characteristic of scientists that they often wonder about questions that others would never think of. It was the Swedish physical chemist Svante August Arrhenius who happened to wonder about the infrared "color" of CO_2.

Arrhenius was a man of broad interests and with an eye for pretty women. He briefly married one of his former students in order to legitimize the birth of their son. Then, eleven years later, he married another lady with whom he remained for the rest of his life. His most consistent love, however, was science.

By 1896, it had been forty years since the development of

the first spectroscopes, and by then the newer laboratory spectrometers were capable of creating quantitative photographic records of otherwise invisible wavelengths. Arrhenius calibrated his apparatus, shined a broad spectrum of light through a gaseous sample of CO_2, exposed a series of photographic plates, then developed them in his darkroom.

The result was that CO_2, as expected, was almost perfectly transparent to visible light and to near-infrared radiation. At slightly longer wavelengths, however, this began to change. First, there were two narrow bands of virtually complete absorption—where, if the human eye were sensitive in that region, the CO_2 would appear essentially black. At even longer wavelengths there appeared a band of partial absorption, followed by another band—this one very broad—that indicated total absorption of the radiation.

Being a competent scientist—in 1903 he would win the Nobel Prize in chemistry for some of his other work—Arrhenius puzzled about his curious finding. Molecular nitrogen and oxygen, being the major constituents of the atmosphere, were known to absorb relatively little infrared radiation. Now, Arrhenius had discovered CO_2 to be a strong absorber of infrared, and its atmospheric concentration—due to global expansions of heavy industry—was surely rising. Might an increase in atmospheric heat–trapping CO_2 have any effect on the future climate of the planet?

Arrhenius did some calculations based on what was then known about the solar spectrum, the composition of Earth's atmosphere, and the laws of radiative heat transfer. In particular, he used the Stefan-Boltzmann Law, which had been well established in laboratories by that time. He was overjoyed by his results: Planet Earth was warming—and there would be no more ice ages!

Arrhenius tried to be as specific as possible. He computed that a doubling of Earth's atmospheric content of CO_2 would lead to an average global temperature increase of 5–6°C. The currently accepted range is 2 to 4°C. He also estimated that such a doubling of CO_2 from human activity would take about a thousand years. In fact, we are already about a third of the way there in just one century.

Meanwhile, Arrhenius reasoned that the corresponding gradual increase in average global temperature would extend growing seasons everywhere—thus providing the means to feed the planet's expanding population long into the future. In other words, he believed sincerely that the human race would be saved by industrialization!

Arrhenius also undertook the reverse computation. If the atmosphere's CO_2 concentration declined by half, he figured, the planet would cool by an average of 4–5°C. This, he speculated, was a plausible explanation for the past ice ages. Although he realized that this computation did not constitute actual proof, he could think of no way to extend the analysis further. So, finding himself at a dead end, Arrhenius moved on to more tractable chemical investigations that gained him greater recognition.

MOVING PLATES

As I touched upon previously, the fossils of warm-weather plants and animals (large reptiles, for instance) are sometimes found in what are, these days, cold climates where they could not possibly survive today. Reasonably for the time, nineteenth-century scientists interpreted such finds as evidence for past fluctuations in global climate. After all, what alternative explanation could there be?

In fact, there was one: continental drift. Interestingly, the idea that the continents are on the move was proposed as early as 1596 and was re-suggested from time to time over the following centuries. Even Benjamin Franklin had speculated about it. In 1912, the geologist and climatologist Alfred Wegener pulled together all the circumstantial evidence he could find (for instance, virtually identical earthworms native to both South Africa and South America) and announced his theory that Earth's landmasses had all once been joined in a single supercontinent. This was later to be named Pangaea, a great mass that Wegener described as being eventually broken apart, the pieces gradually drifting away from one another.

According to Wegener's theory, it should come as no surprise that reptilian fossils are occasionally found in the far north and far south; in the distant past these places were all conjoined in Earth's then-tropical and temperate zones.

At first, most geologists rejected Wegener's theory on the basis that he could propose no plausible mechanism by which whole continents would move. As a result, and until the 1950s, anomalous fossil finds continued to be interpreted solely in terms of global climate change.

Then, around 1960, it became increasingly evident that Wegener had been on the right track after all, and that Earth's crust did indeed consist of a collection of plates that slide here and there, lubricated by the underlying layer of our planet's viscous, semi-molten rock mantle.

The resulting theory was that of "plate tectonics." It offers a powerful explanation for earthquakes and volcanoes, and it is now generally accepted by the scientific community as a true picture of Earth's surface. This development compounded the challenge of unraveling prehistoric climate changes. If a piece of Earth drifts from a spot near the equator to somewhere nearer one of the poles, would one not expect its climate to change drastically on that basis alone? The answer, of course, is yes.

By the time plate tectonics appeared on the scene, however, several new and clever techniques had been developed to chronicle Earth's ancient climate. It was actually becoming possible to separate atmospheric, astronomic, and tectonic factors, insofar as they may have separately driven the climate changes of the past. One result became clear almost immediately: Earth's ancient climate sometimes changed *much* more rapidly than the slow drifts of tectonic plates could possibly account for.

SEEDS AND SEDIMENTS

Learning how to establish the dates of prehistoric events did not come easily. Using just dendrochronology—the counting of tree rings—usually carried a scientist back less than a millennium. Even then, the result was a local rather than a regional or global record. The dateline could be extended to a few thousand years by using overlapping samples of ancient trees preserved in places like Irish peat bogs. Nevertheless, that was about as far as tree rings would take you.

Similarly, layered annual deposits of sediment in certain lake beds could (with difficulty) be counted. This also gave the age of pollen found imbedded in such layers, which then allowed the dating of other regional sites where similar pollen might be found and analyzed microscopically. Such procedures were not only tedious, but their results were also somewhat fuzzy.

This collection of approaches did, however, suggest that the last ice age ended about 10,000 years ago. It also suggested that by 5,000 years ago, parts of Europe were warmer than they are today. For instance, pollen established the age of nuts stored by squirrels—nuts that came from trees that can survive only in warm climates.

Meanwhile, other investigators drew conclusions from fossilized layers of shells in seaside cliffs—which additionally bore evidence of large variations in sea levels over time. Yet again, none of this was particularly accurate quantitatively. All it did was to confirm that several periods of glaciation had occurred in the remote past, at intervals of perhaps 10,000 to 100,000 years.

So why did these early paleoclimatologists simply not date the rocks found at the sites of prehistoric glaciers, using the radiometric procedures developed in the early twentieth century? Quite simply, because that would not have told them much of anything. Some rocks may be quite young—those scoured from the slopes of volcanoes, for instance—while others may be nearly as old as the planet itself. Certainly, in this way, one could come up with some numbers that related to rock samples, but those figures would not and could not necessarily tell you anything about climate change. So the time scales of prehistoric global climate change continued to remain fuzzy—that is, at least, until around 1950.

CELESTIAL WOBBLING

In 130 BCE, on the Aegean island of Rhodes, the astronomer Hipparchus ran into a problem. He had set out to create the most comprehensive atlas of the night sky ever developed, using giant protractors aimed at the heavens to measure the precise positions of about 850 stars (all, of course, visible to the naked eye from that location).

However, when he compared his results to the less-complete reports of previous stargazers spanning several centuries prior, Hipparchus noticed small discrepancies in the data. Most peculiar was the pattern of those apparent errors: They all seemed to be in the same direction, counterclockwise around the North Pole. Not by much, but by enough, and with enough consistency that it occurred to Hipparchus that maybe the night sky isn't as unchanging as he'd assumed it to be. What if those discrepancies weren't mere human errors; what if the heavens actually did wiggle, very slowly, over long periods of time?

Assuming that the observations recorded centuries earlier

were generally accurate, Hipparchus performed some calculations. He concluded that the combined data was consistent with the hypothesis that the whole sky twists slightly as it rotates daily around Earth's axis, and that the whole star pattern returns to its starting point once every 30,000 years.

Regarding that particular number, Hipparchus was amazingly accurate for his time. As for the reason behind the phenomenon, however, he was wrong. It is not the heavens that are "wiggling"—it is actually the axis of our spinning Earth that slowly wobbles in a small circle. The star we identify today as the North Star was not the "north star" in Hipparchus's time. Nor will it be the "north star" a thousand years from now. The phenomenon has since been labeled the "precession of the equinoxes" and the currently accepted value for the period of the process is 25,765 years.

MILANKOVITCH CYCLES

In 1941, a Serbian civil engineer and mathematician, Milutin Milankovic published a 650-page computational analysis in which he proposed an astronomical origin for Earth's ice ages. By that time, astronomers knew that Earth experiences a whole series of small wiggling movements over long periods of time. What, Milankovic asked, was the overall effect of combining them? Would this lead to any broad pattern of increased and decreased sunlight that might drive global climate changes?

Milankovic began with Hipparchus's well-confirmed 25,765-

year precession of the equinoxes. Additionally, he noted (as first analyzed by Johannes Kepler in 1609) that Earth's orbit around the Sun is slightly elliptical rather than circular—and that this ellipse itself rotates (due to interactions with other planets as well as the effects of the solar gravitational gradient). This causes the seasons to shift slightly in several overlapping cycles, whose most prominent periods are 413,000 years, 95,000 years, and 136,000 years.

Simultaneously, in a different display of its gravitational interaction with the Sun, Earth's inclination to its orbital plane fluctuates back and forth, from about 22.1 degrees to 24.5 degrees, over a 41,000-year cycle.

On top of all this, the plane of Earth's orbit around the Sun shifts up and down slightly with a period of about 70,000 years.

Superimposing all of these astronomical cycles (along with a few others), Milankovic concluded that at intervals of about 100,000 years, the incidence of sunlight absorbed by Earth's northern hemisphere reaches a maximum. By extension, at the midpoints of these intervals, that solar flux is at a minimum. So one should expect glaciation and thawing to happen in direct step with these 100,000-year intervals.

Milankovic's ideas attracted a great deal of attention at the time. Even today, they are still mentioned in many college meteorology textbooks—usually with slightly modified spelling under the heading "Milankovitch Cycles." Many researchers still find compelling grains of truth in Milankovic's hypothesis. Others continue to analyze various inconsistencies and unresolved problems with the idea—which, of course, no researchers would ever take the time to do unless they were taking Milankovic's thinking seriously, on some level at least.

So how does Milankovic's astronomical explanation for ice ages stand up today? His proposition has grown a bit shaky, but it has never completely collapsed. Variations in "solar forcing" due to orbital variations may be at least part of the story. But newer data on the dates of the ice ages do not place them quite so neatly into the string of 100,000-year cycles that correlated so well with Milankovic's 1941 analysis. There must be something else going on as well.

OTHER POSSIBILITIES

One alternative hypothesis was more straightforward and went like this: Because it is solar energy that ultimately drives terrestrial weather, any increase or decrease in the Sun's brightness should certainly upset Earth's climate. It could even be the case that our Sun's output varies from time to time, just as some other stars have been observed to do, even on human time scales.

In the early 1600s, various investigators began to use lenses to project images of the Sun onto a screen in a darkened box or onto the wall of a dimmed room. To their surprise (which is what kept them going), they discovered variable patterns of dark spots on the solar surface.

Called "sunspots," these solar blemishes seemed at first to be relatively uncommon—seldom more than a few of them occurring at any one time. Then, around the beginning of the 1700s, the reported numbers of sunspots began to increase significantly.

Simultaneously, the agricultural growing season was getting longer than it had been at any time during the previous century.

Correlations, however, do not prove causality. If sunspot activity increased as growing seasons lengthened, the correspondence in timing might simply have been a coincidence. Or maybe the data itself was just an artifact of improvements in observation techniques. For example, even on a crudely projected image, you may count two or three large sunspots. In a larger and clearer image, you might even identify 10 or 20; and in a still-better image, your count might be as high as 50 or 100. This results from the fact that what others observed and then interpreted as single sunspots on poor projections are, in fact, found on high-quality images to be clusters of spots.

So were there actually fewer sunspots in the 1600s? Some say yes; others say that we simply cannot be sure.

Around 1830, a German amateur astronomer, Heinrich Schwabe, began to keep detailed records of his own sunspot observations. Some twenty years later, in 1851, he combined his two-decade stack of data with that of earlier observers and concluded that sunspot activity ebbs and peaks in a regular pattern that averages about 22 years for a complete cycle.

Then, barely a year later, several other researchers noticed that Schwabe's proposed sunspot cycles correlated quite well with the shifting of the direction of Earth's magnetic north over time. In other words, the Sun seemed to affect Earth not only through a gravitational interaction, but also in other ways.

We now know that the Sun's energy output does indeed increase and decrease by several hundredths of 1 percent on

Schwabe's 22-year cycle—its maximum intensity correlating with the appearance of more sunspots, and its lesser intensity when there are relatively few. Yet as for whether this phenomenon affects our global climate, hundreds of studies have been conducted and none has ever come up with more than the skimpiest of supporting evidence.

Today, the idea that fluctuations in solar output are causing terrestrial climate change is generally discredited. Billions or hundreds of millions of years ago, the Sun's output may indeed have been higher or lower enough to account for a different global climate. But today's solar energy flux varies so minimally that this cannot realistically be considered to be a driving force behind Earth's recent, or ongoing, climate changes.

Our Sun (fortunately for us) appears to be a remarkably stable star. Moreover, current astrophysical calculations suggest that "Old Sol" will continue to be just as stable for the next five billion years or so.

How Matters Stood in 1950

By 1950, a clear picture had emerged, with Earth clearly billions of years old. There was also strong evidence that its climate had fluctuated wildly over at least the latest few hundred million of those years. Furthermore, this had been a driving force behind the evolution of many new life-forms and the demise of others.

The datelines, however—even those of the more recent ice ages—remained somewhat fuzzy. Moreover, neither the variations in Earth's orbit nor the fluctuations in solar intensity seemed sufficient to explain the ice ages or more recent smaller climate fluctuations.

So, at this point, the only remaining factor that seemed to be capable of affecting global climate was the hypothetical alteration of the chemical composition of Earth's atmosphere.

CHAPTER 5

Appliances of the Sciences

If we knew what it was we were doing, it would not be called research, would it?
ALBERT EINSTEIN

With the conclusion of World War II came a considerable reduction in weapons-related research. Many scientists (particularly those affiliated with universities) took delight in the resulting luxury of being able to choose their own topics of investigation. Within the next decade, nearly half of the 16 million former U.S. servicemen and women took advantage of the GI Bill, with a sizeable fraction of them using that benefit to earn college degrees. As a result, by the 1950s, a large cohort of newly minted scientists and engineers was streaming into the U.S. workforce. Soon after, numerous declassified military technologies (e.g., rocket engines and radar) began to find their way into civilian and scientific applications.

The result was a gilded decade of remarkable scientific and technical achievements, not just in the United States, but also in nations around the world. Relevant to the question of climate change, those developments included the likes of:

- improved dating techniques;
- atmospheric CO_2 monitoring;
- development and commercialization of the transistor and transistorized instrumentation;
- launch of the first artificial Earth satellites; and
- advent of proxy global temperature measurements.

In addition, there was a plethora of climatological information gathered during the International Geophysical Year (1957–58).

AN INDEPENDENT REFEREE

Establishing the dates of long-elapsed events had always been a challenge. If the event occurred during historic times, one *might* be lucky enough to find a human account somewhere, but even then the surviving documents were often meager. If the event of interest was prehistoric—as with the ice ages—all one could usually do was to grasp at little snippets of fuzzy evidence and then make an intellectual leap in order to connect the event to the known human calendar of events.

Then in 1949, at the University of Chicago, Willard Libby got a brainstorm while reviewing some pre–World War II scientific literature. His idea was that there might be a method to date the past that was virtually *independent* of human judgment.

Back in 1940, in the early days of the war, two Berkeley scientists—Martin Kamen and Sam Ruben at the University of California Radiation Laboratory—had announced their discovery of carbon-14 (C-14), a radioactive isotope of common carbon (C-12). This was not exactly an Earth-shattering find, for it had long been known that most of the elements have multiple isotopes, i.e., that some of their atomic nuclei contain an extra neutron (or even a deficiency of one or two) that has a slight effect on the atom's weight but does not affect its chemical properties.

Furthermore, it was also already known that some such isotopes were radioactive. In other words, they had unstable nuclei and eventually disintegrated into a different chemical element with different chemical properties.

A known example was that of the potassium isotope K-40, which spontaneously transmutes into isotopes of either argon or calcium (Ar-40 or Ca-40), with a half-life (i.e., the time it takes for half the nuclei to undergo such a change) of 1.25 billion years, and on which basis the lower limit of the age of Earth had already been established. Yet when Kamen and Ruben discovered C-14 in 1940, it seemed a mere curiosity and with no obvious application.

Moving forward a few years to 1949, Willard Libby began to find his interest greatly stimulated by several features of C-14. This isotope was found to occur naturally in small concentrations in Earth's atmosphere, bound up with oxygen just like all of the rest of the atmosphere's CO_2.

Plants, as well as the cells of animals that eat those plants, are unable to distinguish C-14 from the more common carbon, C-12 (or, for that matter, from C-13). Since all life on Earth is made up of carbon (or organic) chemistry, this also meant that every substance of organic origin contains at least a small amount of radioactive carbon-14. Putting it another way, being organic beings, we too are all just a little bit radioactive.

As for *how* radioactive, that was the beautiful part. Everything alive is exactly the same in that respect. Dead organisms, however, are less radioactive. The longer dead, the less so.

Kamen and Ruben's original lab work had established the radioactive half-life of C-14 close to the currently accepted value of 5,730 years. This is the time it takes for half the carbon atoms in a sample of this isotope to transmute into atoms of nitrogen-14 (N-14). Since it is impossible to observe such a long time span directly, what the investigators did instead was to use radiation detectors to count the rate of emission of beta particles (a product of C-14 radioactive decay) from samples that were very carefully weighed. The result was about 14 disintegrations per minute per gram of carbon (expressed as 14 dpm/g). From this, they were able to compute how long they would have needed to wait for half of any collection of carbon-14 atoms to disintegrate. The result, 5,730 years, is called the C-14 half-life.

On this basis, Libby reasoned that it should be possible to reverse the analysis to establish the age of virtually any organic sample simply by extracting a precisely weighed amount of carbon from the artifact, then putting it in an apparatus that counted the emission of beta particles. If, for instance, the rate was 7 dpm/g, then its age was 5,730 years, or one half-life. If the rate was 3.5 dpm/g, then there were only one-fourth as many C-14 atoms remaining, and the sample's age had to be two half-lives, or 11,500 years.

For yet older samples, the declining disintegration counts could be compensated for by using larger samples of material. In this manner, ages of up to about 40,000 years could be measured, with about 100,000 years being the practical limit (always, of course, assuming the availability of a large enough specimen).

There was just one glitch to all of this. If Earth is as old as we think it is, how is it that *any* carbon-14 still exists in the atmosphere? "Surely, it would have all disintegrated by now?" went the question.

Well, yes, it would have—except for the fact that new atoms of the C-14 isotope are continually being generated in the upper atmosphere through a process involving cosmic rays, the charged particles from outer space that were first discovered in balloon experiments in 1912. As a result, the atmospheric concentration of C-14 has remained fairly constant at about one part per trillion for the past 100,000 years or more.

Libby received the 1960 Nobel Prize in chemistry for developing this radiocarbon dating procedure. By then, the technique had been refined considerably. In particular, the minor variations in atmospheric C-14—variations due mainly to solar effects, but also to the above-ground nuclear bomb tests of the 1950s—had been analyzed extensively by comparing carbon-dating with tree-ring data going back a few thousand years.

For dates less than a few thousand years ago, this recalibration made the procedure quite accurate. For older samples, though, the results were slightly fuzzier, but still quite useful. For instance, an artifact with an age of 6,000 years could be dated to within about one century or so—corresponding to an accuracy of about 2 percent. Thus radiocarbon dating soon became an important part of the scientific toolkit for investigating the past.

There is one potential problem, however, that is related directly to our overall topic. This is the fact that it is not possible to carbon-date a glacier directly, because H_2O contains no carbon. But what we *can* date is organic material trapped and preserved within ice or in sediment. This can be either plant matter or animal tissue (Ötzi the Iceman, for instance, in the opening chapter). A good example is the carbon-dating of frozen woolly mammoths found in the frozen tundra. This has established that the species more or less became extinct about 10,000 years ago—although a small population of dwarf mammoths on Wrangel Island in the Arctic Ocean did survive until 4,700 years ago. By then, the world was a much warmer place than it had been for at least 70,000 years, and those last mammoths may have fallen victim to a successful new predator—humans.

THE HOLOCENE EPOCH

Is there an exact date for the end of the last ice age? The answer is yes and no simultaneously.

Because the great continental ice sheets expanded and contracted numerous times over the most recent two million years or so, many geographic locations near their fringes experienced alternating periods of glaciation and thawing. Some of these coincided approximately in tempo with the Milankovitch Cycles, others apparently driven by other agents. For a particular place, such as modern Quebec or Stockholm, one can indeed establish the final date when the ice retreated and never returned. That date usually turns out to be 11,000 years ago, more or less, depending on the specific geographical location where evidence is collected. However, the result is not the same everywhere. In fact, in some places on the planet, the big thaw is yet to happen.

It takes a very long time for a substantial pile of ice to melt completely, particularly when it is several miles thick. (Remember the frozen turkey analogy in Chapter 1.) Therefore, even if it is possible to establish the date when an ice cover retreated from a particular place, it is probably considerably later than the date thawing first began. For example, based on extracting ice cores from deep within existing Greenland glaciers and analyzing trapped bubbles of air, one recent study dates the beginning of the warming trend that initiated the great melting to 11,711 years ago. Although the precision of that result may be somewhat overstated, it is probably a fairly accurate estimate.

As for earlier ice ages, the series of big freezes that came and went hundreds of millions of years to 2.1 billion years ago—and during which some of the intervening periods seem to have left the polar regions completely ice-free—these are episodes that are not particularly relevant to the issue of Earth's current global climate. In those far-off times, our planet's atmosphere was chemically different, and the solar energy flux also had a slightly different value. The evidence most relevant to Earth's current global climate comes from the most recent ice age that occurred some 11,700 years ago. In geological terms, this is the epoch referred to as the Holocene.

However, as intimated above, the last ice age has not yet ended, at least in one sense. After all, the Arctic Ocean is still largely covered by a blanket of sea ice, northern Canada and most of Greenland are ice bound, and virtually all of Antarctica is glaciated. Whilst it is true that the great ice sheet that once covered most of Europe has retreated northward—as has the continental glaciation that once extended into the northern United States—the Arctic and Antarctic ice caps still have not melted away, as they apparently did several times hundreds of millions of years ago. But are they now on their way toward thawing again? Or is the Holocene simply a temporary warm glitch, after which the great glaciations are destined to resume in full force?

The latter view was the predominant scientific opinion during the 1950s and much of the 1960s. The consensus seemed to be that planet Earth was destined to grow colder again.

How Hot and How Cold?

When we use the words "warm" and "cold," we are obviously invoking the concept of temperature. This is a commonplace term, but one that has a somewhat abstract scientific definition that involves the statistics of molecular motion. In general and on average, hot objects contain faster-moving molecules and cold things have slower-moving molecules. This explains, for instance, why hot water evaporates more quickly than cold water, why most substances expand when heated, and why sound travels faster when the temperature is higher

In order to measure temperature, however, we cannot apply the scientific definition directly because it is enormously difficult to gauge the motions of individual molecules. Realistically, all that we can do is to measure some secondary physical property that changes with temperature, e.g., the volume expansion of a liquid, the electrical resistance of a substance, or the emitted infrared radiation. As a result, temperature measurements are seldom as accurate as the scale on the particular thermometer. Go to a rack of thermometers in a department store, for instance, and check whether they are all showing the same temperature. I guarantee that they are not.

Furthermore, in measuring the temperature of anything larger than, say, a cup of coffee, we encounter another type of problem. Take a stroll around your neighborhood and you will notice spots where the air is slightly warmer or cooler, as well as places where the ground is hotter or colder than the air. In light of such observations, does it make sense to talk about the outside temperature at all?

In a way, it does—provided that you can figure out a way to smooth out the spot-to-spot variations. Indeed, such averaging grows more reliable as the size of the region is increased. In addition, there is nothing wrong in doing this. After all, the fundamental definition of temperature itself is statistical.

TEMPERATURE PROXIES

In 1947, in another post–World War II flash of scientific brilliance, it occurred to Harold Urey at the University of Chicago that it might be possible to reconstruct the average temperatures of times long past. His contemporary proxy for those would be the rate of evaporation from the seas.

He chose this phenomenon because in the atmosphere, oxygen exists naturally in three stable (nonradioactive) isotopes: O-16 at 99.762 percent, O-18 at 0.2 percent, and the tiny remainder as O-17.

In the oceans, where one atom of oxygen is present in every water molecule, one might expect the oxygen to exist in these same isotopic ratios. Yet that is not quite the case, partly because the water molecules having an atom of O-16 are slightly lighter than those having an atom of O-18. The lighter molecules evaporate more readily than the heavier ones—and the higher the temperature, the greater the difference in these isotopic concentrations. As to the technical problems of how to sample isotopes from thousands of years ago, Urey proposed that they would show up in the layers of sea sediment.

A few years later, one of Urey's doctoral students, Cesare Emiliani, undertook the task of reconstructing ice-age temperatures by analyzing an actual series of core samples of sea sediments. He concluded that during the peak of the last ice age, the average global temperature was about 7°F (4°C) lower than today; and even when those ice sheets began to recede, the temperature was still slightly lower than today.

When Emiliani published his results in 1955, his findings were met by both applause and rejection. The applause was for such a clever technique; the rejection because sea sediments are such slimy messes, whose oxygen isotopic ratios might easily be affected by factors other than temperature-driven surface evaporation.

What Emiliani's attempt did establish, however, was that isotopic ratios preserved by natural processes might be a way to reconstruct a record of past temperatures. In practical terms, Urey and Emiliani had simply chosen the wrong source of evidence. They should have chosen glacial ice cores rather than sea sediment.

In fact, roughly a decade later, that's exactly what several other research teams started to do. Drilling into Earth's surviving ice sheets, they analyzed the resulting core samples (including the encased bubbles of ancient atmosphere) to learn about past fluctuations in global temperatures.

Monitoring CO_2

In 1954, the participants at an international agricultural science conference in Stockholm concluded that no scientist on Earth understood the details of how CO_2—which is essential for plant growth—circulates through the atmosphere. In an attempt to remedy this dearth of knowledge, a team of Scandinavian researchers set up a network of fifteen stations in order to monitor atmospheric CO_2.

However, as the results began to flow in over the following months, the numbers were disappointingly inconsistent. It later turned out that the equipment used at the stations had been detecting highly localized CO_2 variations—the levels having been affected by factors such as brushfires, decaying vegetation, industrial activity, and variations in wind direction. Luckily, it had not been a very expensive experiment, and so little was lost financially.

A few years later, the agriculturally-based question about CO_2 would be answered via other investigative techniques—including the use of radioisotopic tracers.

KEELING AND HIS CURVE

Charles David Keeling, a postdoctoral fellow in geochemistry at Caltech, had hoped for a different outcome to the above investigation. He, too, was interested in carbon dioxide, but not for reasons relating to plant respiration. Rather, he was curious about the effects of CO_2 on climate, believing it to be involved in some way, just as Arrhenius had suggested more than a half century earlier.

When planning began for the International Geophysical Year (IGY), scheduled to begin in July of 1957 and initially involving 46 countries (eventually expanding to 67), Keeling proposed that the global research program include installing state-of-the-art CO_2 monitors in Antarctica and on Mauna Loa, Hawaii. The objective would be to create a precise baseline for comparing any and all future measurements of the atmosphere's CO_2 content. Each of the geographical locations was chosen for its remoteness, where the measurements would be unaffected by anything untoward and thereby represent accurately an overall average indication of global conditions. All this was meant to confirm that the data was both consistent and therefore truly global.

It worked—and Keeling indeed got his baseline. Then, in December of 1958, the IGY came to an end, and with that termination also came the end of the funding for Keeling's atmospheric monitoring program.

At this juncture, many people would simply have thought, "Grin and bear it and move on." This might well have been the case, were it not for a significant coincidence. It just so happened that the funding lapse occurred right at a time that the data started to hint—albeit as a bare whisper—that atmospheric CO_2 might be increasing. Any detectable trend in just a year and a half was hard to identify—unless the atmospheric monitoring program was extended.

For the next several years, Keeling spent more of his time chasing funding than in actual science. Unfortunately, long-term investigations (sometimes called longitudinal studies) are of meager interest to bureaucrats and agencies that need to report their results annually. So Keeling had to abandon any effort to keep going with the extremely expensive Antarctica monitoring. On the other hand, the Hawaii measurements alone might have continued on only about $100,000 per year.

With this in mind, Keeling did everything possible to pare down his costs. In the spring of 1964, his spectrophotometer failed and it took several months before it was fixed and up and running again—the irony being that with adequate funding, it would have been functional again within a week or two. The resulting hiatus appears in Keeling's original data, but is usually smoothed over in more recent reports.

Keeling's funding struggles continued for a total of about 18 years, until the U.S. National Oceanic and Atmospheric Administration (NOAA) rescued the program in 1973 by taking over the operation and maintenance of the Mauna Loa Observatory.

By 1976, there was no longer any question that the CO_2 levels in the atmosphere were increasing. Several scientists even began warning that this could lead to global climate change in the near future. At that time, however, few were listening.

The CO_2 monitoring continues to this day. Figure 4 shows the data trend starting in 1959, when the average CO_2 concentration was 315.98 parts per million (plus or minus 0.12 ppm), until 2008 (49 years later), when it averaged 385.57 ppm.

Stated another way, the increase over the past half century has been 22 percent. Thus, discounting the small squiggle due to seasonal plant growth, the entire trend has been upward.

How do these numbers compare with the atmospheric CO_2 content of earlier times? It turns out that today's values are higher than those found in any ice-core data. During the early Holocene—when the great ice sheets first began to retreat some 11,700 years ago—atmospheric CO_2 peaked at about 350 ppm. Soon after that, the concentration declined and settled down within the range of 275–285 ppm. There it remained for about the next 11,000 years. Then, in the early 1800s, it gradually began to rise again. In 1988, it once again reached 350 ppm,

Carbon Dioxide, ppm

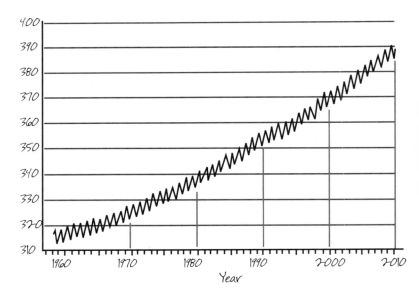

FIGURE 4: ATMOSPHERIC CARBON DIOXIDE CONCENTRATIONS
Atmospheric carbon dioxide (CO_2) concentrations 1959 through 2010 expressed as
parts per million (ppm) and as measured at Mauna Loa, Hawaii. The small annual
fluctuations are due to seasonal plant growth in the northern hemisphere; growing
vegetation absorbs CO_2. (Data from NOAA.)

the same level as when the continental glaciers began to retreat. And, yes, by that date it was also becoming apparent that much of the planet—particularly in the northern hemisphere—was warming. Now, at the time of writing, it is at 386 ppm—higher than at any time within the past several hundred thousand years.

VIEWS FROM ABOVE

The flurry of post–World War II advances in Earth science could not have occurred without sophisticated measuring instruments. The major boon was the 1947 invention of the transistor, which was more rugged, reliable, and energy-efficient than the earlier vacuum tubes. By 1954, the first commercial transistor radios were being mass-produced. At about the same time, transistors began to appear in a wide variety of other electronic devices—including scientific instrumentation.

This timing meshed perfectly with another global technological development: the launching of the first artificial Earth satellites, beginning in late 1957. To withstand the inertial forces of the launch, the onboard electronics had to be extremely rugged. To operate for any significant length of time, the instruments not only had to be reliable, but also energy efficient. The early American and Russian space programs would never have gone very far without the transistor.

In 1960, satellites began monitoring Earth's weather from space. At first, the results were just an unreliable curiosity. Those early satellites had typical lifetimes of less than a year, and months would typically elapse between launches. Good engi-

neers, however, are adept at learning from their failures, and they soon extended the lifetimes of their space-bound instrument packages. By the end of the decade, new weather satellites were being launched more frequently than the old ones were failing.

CATASTROPHIC CAMILLE

In 1969, for the first time ever, a hurricane was monitored from space continuously from its birth through its death. This was no ordinary hurricane. It was Camille, which battered the Mississippi Gulf Coast with record-setting sustained wind speeds as high as 201 miles per hour (based on extrapolations from hurricane-hunter aircraft instrumentation).

The coastal devastation shocked the entire United States, but Camille still was not out of steam when she made landfall. Most of the 108 billion tons of moisture the hurricane had vacuumed from the Gulf were still in her cloud tops. Camille tracked north over Mississippi, crossed Tennessee, and turned eastward in Kentucky. Then, over sparsely-populated Nelson County, Virginia, her moist remnants collided with a cold front—and it rained. In fact, it rained a lot, with as much as 46 inches (117 cm) in about six hours in Huffman's Hollow. The resulting flood was the worst in that region in 10,000 years, and it claimed at least 125 lives.

So what did those "newfangled" satellite images do to warn the prospective victims about the imminent disaster? Virtually nothing. The photos were relayed to the Washington headquarters of the U.S. Weather Service, where nobody quite knew what to do with them. A few of the images did reach the National

Hurricane Center, but after Camille made landfall, even those belated forwardings ended.

CATEGORIZING HURRICANES

In Camille's tragic aftermath, Robert Simpson, then the director of the National Hurricane Center, tramped through the resulting coastal rubble. Hurricanes were obviously not equal in their character, he concluded, and there needed to be a category system to rate the human risks. Thus was born the now-familiar Saffir-Simpson Hurricane Wind Scale, with its hurricane categories ranging from 1 through 5.

Retrospectively, Camille was ranked as a category 5 hurricane, the very worst. Only two others of category 5 have made landfall in the U.S. since records have been kept. One was an unnamed storm that struck the Florida Keys in 1935, and the other was Hurricane Andrew in 1992—which was only designated as a category 5 some 10 years after the event.

Camille clearly demonstrated the value of monitoring Earth from space. Soon a wide variety of precision measurements were being made through satellite telemetry—including surface temperatures, changes in Earth's magnetic field, changes in sea levels, solar phenomena, distributions of sea ice—and the list goes on. Today, if scientists need longitudinal data about such variables, they usually find that several decades' worth have now been archived.

CHAPTER 6

Warmer or Colder? Wetter or Drier?

When you're hot, you're hot;
When you're not, you're not.
FLIP WILSON

The hottest day in the record books was September 13, 1922, when the temperature in Al Aziziyah, Libya, skyrocketed to 136°F (57.8°C). The previous world record was set on July 10, 1913, when the mercury climbed to 134°F (56.7°C) in Death Valley, California. At the other extreme, the lowest recorded temperature was -128.6°F (-89.2C) on July 21, 1983, at Vostok Station, Antarctica. Surface temperatures on Earth have thus been observed to span a range of 265°F (147°C)—much greater than the range on our sweltering cousin, Venus, but slightly less than that on frigid Mars.

Record temperatures certainly tell us something about Earth's climate. But do they reveal anything about global climate change? Alas, not one bit. If they did, it would be like saying that the outstanding feats of the strongest man on Earth prove that the rest of us are getting stronger. Or, conversely, arguing that we're all getting weaker, because that strongman accomplished those feats 60 years ago, and nobody has surpassed them since.

There are several problems with placing a great deal of emphasis on meteorological extremes. First, the data does not extend very far back. Out of the tens of thousands of towns and cities in the United States, for instance, there are fewer than 150 for which there exists any meteorological data dating before 1870; and for most of the rest of the world, the database is even sparser. In some underdeveloped countries, and in uninhabited regions, such longitudinal data can even be nonexistent. As for the oceans, which cover a full 70 percent of our planet, temperature data only goes back to around the mid-1980s; and precipitation data still remains intermittent at best.

Second, even in places where the databases do stretch back more than a century—for the world's major capital cities, for instance—new records are still certain to be set from time to time.

At any location, in the first year of meteorological data keeping, every single measurement sets a record for that day of the year. Then, in the second year, about half of the records set new records, and so forth. Even after a century, the statistical expectation is that a couple of days a year will still experience a record high temperature, a record low temperature, or a record rainfall for that date. However, outcomes like these have little to do with the realities of climate—because it is perfectly possible that even more extreme conditions occurred the very year *before* the data-taking began, almost certainly within the preceding century.

This raises the following dilemma: If we are unable to determine the normal long-term temperature, then how can we identify whether or not it is changing?

In practice, there is only one feasible strategy to determine any global change, and that is to apply the law of averages—or, more accurately, what is termed the "law of large numbers." This involves combining a whole lot of data from many, many places, while applying some statistical adjustments to ensure that there is no emphasis favoring one geographical region over another. Then, in repeating this exact same procedure from year to year, one can hope to determine whether or not there is any perceivable trend. Is the average global temperature steady? Is it decreasing? Or might it be increasing?

One major advantage of this statistical procedure is that we do not need to proclaim the accuracy of the *actual* average global temperature. In fact, different sources quote that figure as between about 55°F (13°C) and 63°F (17°C)—with most of the differences resulting from the examination of different time-spans. The actual number, however, is of no real consequence. What truly matters is whether or not that number is changing, and if so, in which direction.

The following analogy illustrates this concept quite well. Imagine that there is a lily pad in the lake next to a boat dock. Being curious about whether or not it is growing, we can lean over it, extend an arm as high as possible, and slowly dribble a half-cup of dry rice over the plant. We then count the number of rice grains that landed on the leaves, say, 138. Several days later, we repeat the procedure and find that it is 149. Of course, in no way does this prove anything; the change could simply be a statistical fluctuation. So we do the same thing again the next day and find that it is now 155. And so on, until the evidence eventually accumulates that the leaves are indeed growing, simply because they're catching more rice. Not only that, but we can even calculate that they have grown approximately 12 percent over four days. As for the *actual* surface area of the leaves, however, we still have no idea. After all, we have not measured the area, but merely the *change* in the leaf's area.

In 1987, the Goddard Institute for Space Sciences (GISS) published a statistical analysis of temperature data from hundreds of ground-based meteorological stations covering the period 1880–1985. The conclusion was that the average global temperature had indeed risen during the previous century by some 0.9°–1.3°F (0.5°–0.7°C).

However, that initial report failed to attract much journalistic or media attention. Apparently, the findings were not sufficiently sensational. After all, if you change the thermostat in your home by one degree, you may not even notice the difference. The same argument seemed to extend to matters global. If our planet has taken a whole century to warm up just by one degree, then how could that possibly be any cause for concern? Erroneous perhaps, but the media failed to connect the temperature increase to the fact that Earth's ice is melting. Furthermore, scientists in 1987 failed to point out such an oversight, probably because the link is not a particularly easy one to explain.

LOCAL AND GLOBAL

One effect of averaging a large pool of data is that it smooths out the random fluctuations. At the same time, of course, averaging also reduces the size of the changes that are quite real. For instance, suppose that your hometown experiences an unusually warm February, with temperatures averaging 10 degrees above normal. If the rest of the year is normal, then the *average* annual temperature is only 0.8°F higher than normal, which may not sound like much. Yet that warm February can easily cause unexpected melting of snow and flooding, both of which *are* quite notable effects.

Furthermore, suppose that we average your town's February temperatures not only over time, but also over a larger geographical region. Say that this involves two other towns, both of which experience normal Februaries. Now the regional temperature increase is reduced to just 0.3°F. The increase has shrunk, but that still failed to prevent the local snow from melting or the floods from arriving.

In fact, this is essentially the disconnect that occurred when the GISS scientists concluded, in 1987, that the average global temperature had increased by about 1°F over the previous century. They were not saying that the whole planet had warmed up uniformly by that amount. On the contrary, some places actually had become slightly colder, while others grew noticeably warmer. Moreover, the places where the temperature increases were most prominent were in the respective summers in the Arctic and a few parts of the Antarctic.

GLOBAL TEMPERATURE MEASUREMENT

Regardless of what the 1987 GISS global temperature report did not accomplish, it *did* alert scientists to the fact that global temperature needs to be monitored continuously.

Combining data from hundreds of ground observatories, however, has its drawbacks. First of all, temperatures collected that way are inherently "noisy." The ground absorbs and releases heat at rates that are affected by numerous variables: clouds, vegetation (which in turn is linked to variations in precipitation), wind (which may spin dust into the atmosphere), the presence or absence of snow cover, whether or not that snow is dirty, and so on. Although the local temperature figures may be correct, their links to global climate are problematic.

The second shortcoming of ground-station temperature data is that most of Earth's surface is liquid, not solid.

Both drawbacks can be overcome by looking at the world's seas rather than its lands. In fact, sea temperature data had already been recorded for many years, but not very comprehensively or consistently. Seagoing vessels had long ago started monitoring the sea temperature in their cooling water intakes. Other temperature data, particularly near coastlines, was recorded by buoys originally tethered to the seafloor in order to record tidal variations.

In the mid-1980s, however, there were no uniform standards for making such measurements. In particular, the results did not all correspond to the same depth of water. Further, in some oceans, a relatively small difference in depth alone can account for a big difference in temperature—much greater, in fact, than the fractional-degree global changes that scientists sought to detect.

The solution was clear: Use satellites to measure sea temperatures by detecting the infrared signature of the ocean surface. All objects, unless they are at a temperature of absolute zero, emit electromagnetic waves, and the warmer the body, the shorter the wavelength of that radiation. For the surface temperatures on Earth, the emitted wavelengths fall into the short-microwave and long-infrared regions of the electromagnetic spectrum. As a result, a suitably designed orbiting radiometer can scan the seas as well as the lands and create a temperature profile of Earth's entire surface.

As for a consistent depth, that is now no longer a problem. Radiometric instruments measure the temperature of only the top fraction of a millimeter, the so-called "skin temperature."

The resulting data shows that since around 1990, the average global temperature has increased by another 0.9°F (0.5°C). Again, this is an average; there are some spots in the Pacific, central Canada, southeast South America, and off the coast of Antarctica where the average temperature has actually dropped by as much as 2.7°F (1.5°C) over the same period. But most of the world has become warmer, particularly Scandinavia and Siberia —and at points within the Arctic Circle, the two-decade rise has been a huge 3.6°F (2.0°C) on average. No wonder the Arctic ice is retreating more and more each summer.

Figure 5 shows how the average global temperature has fluctuated from the year 1880 to the present. The baseline on this

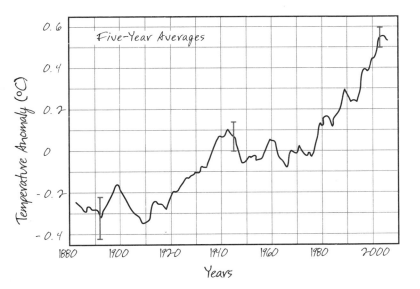

FIGURE 5: GLOBAL TEMPERATURE ANOMALIES
Global temperature anomalies 1880–2008, relative to an average 1951–1980 baseline.
The two decades between 1980–2010 have been the warmest on record. (Source:
NASA/GISS.)

graph is the 30-year average from 1951 to 1980, and the points on the graph are the temperature anomalies relative to this baseline. Thus, the years prior to 1937 were generally cooler (globally speaking), whereas since 1988, as shown on this graph, the average global temperature has trended significantly higher than the baseline average.

As mentioned, however, the world is not warming uniformly. Most of the increase has taken place outside of the equatorial belt, at latitudes north of the Tropic of Cancer, where, as shown in Figure 6, the average anomaly is now about 1.8°F (1.0°C). Meanwhile, at latitudes south of the Tropic of Capricorn, the average anomaly is only about 0.5°F (0.3°C)—which, of course, is still positive. Is it a coincidence that most of the world's population, as well as its industrial activity, reside at the more northern latitudes where there has been a greater temperature increase? Maybe it is; but probably not.

As stated in the Preface (Figures 3 and 4 notwithstanding) it is probably unfortunate that this phenomenon was initially labeled as "global warming." We need to remember that these analyses of average global temperature change are simply an analytical tool, whose purpose is to detect climate trends that would otherwise be lost in the "noise."

Temperature itself is not the whole issue. What we need to be concerned about are a series of ancillary effects: the melting of polar and glacial ice, sea-level rise, possible changes in weather patterns (including extreme weather), effects on agriculture, and effects on flora and fauna.

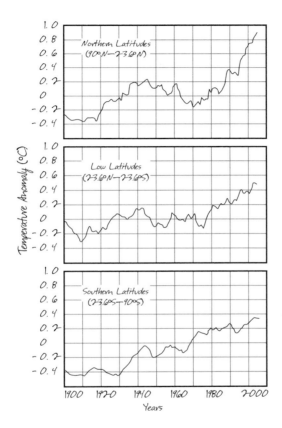

FIGURE 6: TEMPERATURE ANOMALIES BY LATITUDE
Average temperature anomalies vary with different bands of latitude. The greatest warming has taken place at northern latitudes, which is home to most of Earth's population and most of its industrial activity. The baseline in each band is the average 1951–1980 temperature in that band. (Source: NASA/GISS.)

Oh, and yes . . . we need also to worry about whether or not such effects will be reversible.

MORE OR LESS RAIN?

What about precipitation? Is that also changing? One expects that it is—on the simple basis that warmer oceans evaporate more rapidly, which means more moisture circulating in the atmosphere. On the other hand, any overall increase in precipitation could easily fall right back on those same oceans, making the entire process a washout. Or maybe the weather patterns on a warmer Earth will result in less rain and snow falling on land and more on the seas, or vice versa?

It may seem that this should be an easy scientific question to answer. After all, it is surely a great deal easier to measure rainfall than to measure temperature. Alas, this is only the case on land, for out at sea, it is almost impossible to tell how much rain falls on one spot as opposed to another.

Even on land, where accurate data is available, precipitation trends remains difficult to decipher. Rainfall and snowfall can be wildly variable from one year to the next—much more variable than temperatures. As an example, I consulted the precipitation data for one city, Tallahassee, for the year 2008. Table III shows what I found.

Notice that the actual rainfall was close to the long-term average in only one of 12 months: April. Notice also that if it had not been for a record-setting rainy August that year, the overall rainfall deficit would have been at least a foot rather than the

Table III: Monthly precipitation, in inches,
in Tallahassee in 2008, compared to monthly
averages and monthly records (since 1885).
Most other regions experience similarly large
variations in precipitation.

Month	2008	Avg. since 1885	Record/Year
January	3.53	5.36	18.94/1991
February	8.31	4.63	12.22/1914
March	2.41	6.47	16.48/1948
April	3.74	3.59	13.13/1973
May	3.22	4.95	12.36/1890
June	5.77	6.92	17.41/1989
July	4.27	8.04	20.12/1964
August	16.52	7.03	16.52/2008
September	1.29	5.01	23.85/1924
October	4.26	3.25	12.27/1959
November	5.68	3.86	12.64/1947
December	1.39	4.10	12.78/1907
Total	60.39	63.21	

actual three inches. But I chose Tallahassee only as a convenient example (since I live close by). Check the corresponding data for your own locality and you will surely find similarly large variations in precipitation.

On the one hand, data like this is good news, because it shows that we humans are used to coping with much greater fluctuations in precipitation than in temperature. On the other hand, such data is also sobering, because it tells us what is empirically possible: that it could easily get a great deal wetter (or dryer, for that matter).

COMMUNICATIONS BREAKDOWN

Scientists, like most other folks, interact mostly with people who closely share their interests. Climatologists, for instance, communicate regularly with other climatologists, but less so with physicists, and generally even less with planetary scientists. A consequence of this can be that when scientists speak without full knowledge of what their colleagues in the overlapping disciplines are studying, they can sometimes confuse everyone— including one another.

This was very much the case in the 1970s–1980s. Even as the evidence was accumulating that the world was warming, there were dire warnings from some scientists of a possible return of the ice ages.

What actually happened during that 1970s–1980s period has been widely misinterpreted. The truth is that there was never any serious scientific conflict about the basic climatological facts. All agreed that Earth's ice was melting, CO_2 levels were rising, and average global temperatures seemed to be increasing. Yes, there were quibbles about how much, how fast, when, where, and even why, but not about the basic factual scenario. Those fundamental facts were pretty clear to most of the scientific community by the early 1980s.

Meanwhile, however, data from ice cores drilled in Greenland began to suggest that the beginning of the Holocene period (roughly 11,000 years ago) might not actually have been the end of the last ice age. Previous ice ages were all punctuated by warm interglacial periods that often lasted thousands of years. Based on those patterns extending back millions of years, there was a possibility that Earth's current temperate climate is no more than an interglacial fluctuation. If so, then Earth's temperatures will probably plummet again and the great ice sheets will return— not tomorrow, of course, but eventually.

This kind of reasoning was based on long, slow patterns of past climatological shift and was quite different from reasoning based upon a mere century or so of meteorological data. Also, both sets of conclusions could actually be correct. In the short term, perhaps the prognosis was indeed for more warming before the return of a big freeze. On the other hand, maybe human-generated CO_2 was overriding the eons-long pattern of ebbs and growths of the great ice sheets—and in such a way that the planet's climate was *irreversibly* warming.

Curiously, no single journalist seems to have attempted to combine both stories, let alone clarify the nuances. Some reported that a return of the great ice sheets was imminent. Others wrote that the planet was thawing. As for the general public, most simply shrugged in confusion.

NUCLEAR WINTERS AND VOLCANIC BLACKOUTS

Compounding the puzzlement, in 1983 Carl Sagan and several of his colleagues published their theory of a "nuclear winter." Sagan was a planetary scientist affiliated with Cornell University and the Jet Propulsion Laboratory in California and the author of several highly acclaimed books popularizing science, including *Broca's Brain, The Dragons of Eden,* and *Cosmos.* Pretty much everything he published drew instant attention.

One of Sagan's major fears was the prospect of an exchange of nuclear weapons between the United States and (what was at that time) the Soviet Union. In 1980, the two nations had a combined total of more than 20,000 nuclear warheads aimed at each other in a nuclear standoff where both sides were actually running out of targets. Sagan wondered about the prospective climatological aftermath of even a limited nuclear weapons exchange.

There were a couple of historical precedents to guide Sagan's thinking. One was the devastating 1883 explosion of the Indonesian volcano Krakatoa in what was then the Dutch East Indies. Krakatoa ejected vast volumes of volcanic ash and many tons of sulfur oxides with such force that they entered Earth's stratosphere. Within a few months, that eruption cloud blanketed most of the northern hemisphere.

The post-Krakatoa sunsets were, by all accounts, magnificent. Yet in much of Europe and the northern half of the United States, the next growing season was cut short by unseasonably early frosts. Temperature data from 1884 is rather limited, so it is difficult to gauge the actual global temperature change. Nevertheless, it would appear that the Krakatoa eruption suppressed the average temperature in the northern hemisphere by at least 1°F for more than one year. The reason for this was that the high-altitude eruption clouds increased Earth's albedo.

Volcanic explosions on the scale of Krakatoa are relatively

rare. Yet, in 1815, there had been an even bigger one. In fact, it was about 10 times as large in terms of the total material ejected. That volcano was Tambora, also in the then Dutch East Indies. The imprint of its acid fallout is still to be found in the Greenland ice cores, confirming that its effects were worldwide.

The following year, 1816, was the so-called "year without a summer." Temperatures recorded at Yale University in Connecticut averaged at least 7°F (4°C) lower than normal. In June, it snowed as far south as Massachusetts. A series of frosts in June, July, and August caused crop failures throughout the whole of the northeastern states. Compounding matters, the weather was uncharacteristically dry. Much of Europe experienced a similarly cold summer and its consequent crop failures. Grain prices tripled. Food riots took place in France, the Netherlands, and Switzerland. In Switzerland, the authorities issued instructions on avoiding eating poisonous plants (even though such species are hardly plentiful there). Cat and dog populations plunged. Ireland suffered a terrible famine. All of these events, apparently, occurred because of a volcanic eruption on the other side of the world.

But back to Sagan and his nuclear winter. It was clear to him that when a sufficient number of volcanic particulates and enough sulfur oxides are ejected high enough in the atmosphere, they reduce the average temperature of the entire hemisphere. A nuclear war would surely do the same.

Sagan and his team developed a computer model and ran it under a series of alternative scenarios. The results showed that if several dozen nuclear bombs struck cities on both sides of any

such conflict, all survivors in the northern hemisphere would experience many months of freezing weather. If that nuclear exchange were more extensive, then the resulting nuclear winter would be subfreezing and would last several years or more. In other words, it would be long enough and cold enough for much, even most, plant and animal life to die off.

Those 1983 results had nothing to do with the other ongoing research on global climate change. Yet, unfortunately, they added to the public confusion about warming, cooling, warming, cooling, and so on.

CHAPTER 7

≋

Ups and Downs, Ins and Outs

I must go down to the seas again, to the lonely sea and the sky,
And all I ask is a tall ship and a star to steer her by . . .
JOHN MASEFIELD, "SEA FEVER"

The Native Americans knew better than to build their permanent settlements directly on the coasts. Oceans can become violent with meager warning, and when they do, they can wreak terrible devastation. When they did venture to the beaches to fish or gather mollusks, those peoples promptly carried their sea-harvests safely inland.

In the rest of the world, however, numerous early peoples were audacious enough to take their chances with the seas. In many cases, the gamble of building on a shoreline paid off, and seaports around the globe bustled with commerce even in ancient times. Never mind that every once in a while the seas would obliterate one of those coastal settlements.

The United Nations reports that approximately 700 million humans now live near the continental coastlines at elevations of less than 33 feet (10 m). The technical term for such places is "low elevation coastal zones" (LECZs). Remarkably, some 65 percent of the world's cities with populations greater than 5 million lie at least partially in an LECZ. In Bangladesh—the most densely populated nation in the world—fully 46 percent of the population, or 63 million people, live in an LECZ. In the United States, the portion is about 8 percent, which still amounts to at least 23 million people.

While the oceans will not rise 33 feet anytime soon, they are rising. As they do, everyone inhabiting an LECZ becomes increasingly vulnerable to storm surges, sea encroachment, and coastal erosion. Meanwhile, almost everywhere on the globe, the populations in LECZs are growing. Only a minor portion of this growth relates to local birth rates. The rest results from immigration.

COASTAL POPULATION MAGNETS

Humans have always thronged to places of economic opportunity, be they real or merely perceived (e.g., the nineteenth-century gold rushes). Further, wherever money is to be made from one particular activity, there are always spin-offs into other sectors of the regional economy: manufacturing, the arts, education, research, communication, transportation, and entertainment . . . the list goes on. Spurred initially by maritime commerce, the world's big coastal cities have now become people-magnets for a wide variety of reasons—including tourism.

Middle-class tourism is a relatively recent phenomenon, which for the most part dates back only to the post–World War II global economic boom. By the late twentieth century, anyone who invested in a motel on a beach could be assured that customers would come. Before long, such developments became increasingly elaborate, which in turn created local job opportunities, which in turn led the nearby communities to expand further, and so on.

As just one example, the city of Cancún, Mexico, was nonexistent as recently as 1970. At that time, only three people lived on the peninsula and just 117 in the nearby fishing village of Puerto Juárez. Today, Cancún has a permanent population of 600,000, about 150 hotels with more than 24,000 rooms, at least 380 restaurants, and about 200 flights daily from its modern airport. Thus it is not just Cancún's beaches that are at risk from rising seas and the prospect of extreme weather—it is the entire regional economy.

Cancún is not in any way unique, as there are thousands of resort cities and towns similarly vulnerable around the globe—as are thousands of coastal retirement communities.

TIDES, LEVELS, AND TROUBLES

In the century prior to 1993, the rate of sea-level rise averaged about 0.7 inches (1.8 cm) per year. Since then, orbital satellites have measured the annual rate at between 0.11–0.12 inches (2.8–3.1 mm)—which is equivalent to about an inch every eight or nine years.

The seemingly simple concept of sea level, however, is a slippery one, especially when we try to define it precisely. Stand at a wharf and watch the water; clearly, its level is anything but constant. In the short term—seconds—waves ripple the surface. Then over the course of a day, the average level rises and falls twice—a consequence of the gravitational interaction between the sea and the Moon. These tides, in turn, are not the same from day to day or month to month; sometimes their highs are higher and their lows are lower, depending on the season and the relative positions of the Sun and the Moon. The weather also has an effect, and sometimes a big one.

The notion of sea level is based on an idealized calm sea— one with no waves or tides, with no wind blowing over it, and with the surface experiencing a standard barometric pressure of 29.9 inches of mercury at a temperature of 68°F (which is equivalent to 760 mm of mercury at 20°C).

Nevertheless, just as we find the concept of a straight line useful—despite the real-life impossibility of such a geometrical abstraction—so does the concept of sea level have its usefulness. By knowing where the sea ought to be, scientists can describe separately the deviational effects of waves, tides, temperature changes, influxes of fresh water, storm surges, and even tsunamis. Such information allows engineers and public planners to make informed decisions about the design and placement of shoreline structures from wharves to roads and homes and businesses. It enables emergency managers to define and prepare for coastal storm worst-case scenarios. In addition, over the long term—decades or more—it informs scientists about the effects of global climate change.

The current numbers—a sea-level rise of 0.125 inches (3.2 mm)—may not sound like much. Eventually, however, those numbers will add up; and as they do so, the consequences may be grim for millions of coastal dwellers. Moreover, the economic and environmental ripple effects are likely to affect virtually everyone on the planet.

RISING SEAS

Why the seas are rising is relatively straightforward.

First comes the process of thermal expansion. As the oceans become warmer, they swell slightly, just as most things do indeed expand when heated. Although this alone is not a huge effect, it is likely to account for a rise of more than a foot (30 cm) over the next century. Even though this thermal expansion is actually moderated slightly by the initial contraction of meltwater as it warms up from 32°F to 39°F (0°C to 4°C), thereafter it experiences pure expansion with rising temperatures.

As for the melting of polar ice caps, that produces a much smaller effect on sea levels. In fact, if the Arctic Ocean were freshwater rather than saltwater, the melting sea ice would have no effect on sea level at all—the ice there would already be displacing its own volume. However, this is not quite what is happening. Ocean water is slightly denser than the fresh meltwater that comes from the floating ice. As a result, every thawed chunk of floating ice adds about 3 percent of its total volume to the net volume of the oceans. Although this is a small amount, when it is converted to a sea-level rise, this is an increase and not a decrease.

It is quite another matter, however, with continental ice sheets, glaciers, snowpacks, and permafrost. These frozen masses have sequestered and stored up water on *land* for thousands, even hundreds of thousands, of years. As they melt, the runoff water ends up in the seas within a matter of days or, at most, weeks. There is enough ice covering Antarctica and Greenland that if most of it thawed, the oceans would rise a huge 230 feet (70 m) or so—15 percent of this coming from the Greenland ice sheet and 200 feet (60 m) from Antarctica. Even though this may sound like an outlandish scenario, it has actually happened several times during Earth's prehistory.

Fossils of ancient seashells and other sea creatures are often found at elevations far above our present sea level. Samples from boreholes have also been chemically analyzed for evidence that various sites were submerged under saltwater in the past. Numerous researchers have combined such evidence from around the world to reconstruct prehistoric sea levels. While there is never full agreement on the actual numbers, most paleoclimatologists do agree on some general patterns.

During the last ice age, for instance, the seas were about 300 feet (90 m) lower than they are today. Some researchers say as much as 400 feet (120 m) lower. The Bering Strait, between Alaska and Siberia and 53 miles (85 km) wide at its narrowest, is today only about 160 feet (49 m) deep. Thus, whether the seas were once lower by 300 or 400 feet makes little difference because it required only a drop of some 160 feet to create a land bridge between Asia and North America. Similarly, regardless of whether that isthmus lasted tens of thousands, or hundreds of thousands, of years, there was more than ample time for many new species—including humans—to migrate from one continent to the other.

Peering back further into the past—to epochs where nothing resembling a human walked the Earth—sea levels varied even more dramatically. Around 35 million years ago, normal sea level may have been as much as 660 feet (200 m) higher than today—although some sources say only around 330 feet (100 m) higher. In the early Eocene epoch—about 55 million years ago, when Earth seems to have been about 11°F (6°C) warmer than today—sea level was about 230 feet (70 m) higher than now. At the time of the demise of the dinosaurs (65 to 70 million years ago), the seas seem to have been higher still—possibly 550 feet (168 m) higher. Some 200 million years ago, however, the oceans were about at today's level. Further back still, between 250 and 300 million years ago, sea level was as much as 100 feet (30 m) lower than today; and so on.

Why is there such the fuzziness in the numbers? Some of the lack of precision has to do with the difficulties of establishing dates that go back tens of millions of years. Other associated reasons arise from the fact that Earth's crust is not static over geological time periods. The huge surface plates not only drift, but they also heave, twist, and sometimes subside. Finding a sea fossil high on a hill does not guarantee that the ancient organism died at that exact elevation. Over the course of millions of years, a local shift of 100 feet (30 m) upward or downward is quite possible. Consequently, it comes as no surprise to scientists that data collected in different parts of the world do not match perfectly.

Overall, however, there is little doubt that sea level has fluctuated drastically over geological time spans. And it will surely change again. The questions facing us today are which direction will it change, how fast, and with what human (and other) impacts? Plus, of course, might we be bringing on the changes ourselves?

How's the Weather?

On the eve of World War II, military planners in the United States, Britain, Japan, Germany, and the other major powers became increasingly concerned about weather forecasting. Meteorological predictions had never been particularly accurate even in these countries. Now there was the prospect of deploying ships and troops in far-off places, locations where the weather would be an even bigger unknown. As every military leader after Napoleon well understood, the success of whole military campaigns could stand or fall on the weather.

In those days, the only practical way to predict weather was to glean patterns from the historical data and assume that Mother Nature would continue to follow those same scripts. In the United States, for example, forecasters noticed that every time a moist warm front surged north from the Gulf of Mexico and encountered a cold air mass sweeping south from Canada, the collision generated thunderstorms—and sometimes even tornadoes. Thus, the next time such a collision appeared imminent, it made sense to predict stormy weather and declare a tornado watch for the collision region.

Most weather patterns, however, were considerably more subtle than head-on crashes between air masses of drastically different temperatures and moisture content. As a result, the accuracy of most daily forecasts was therefore quite low in those days. So low, in fact, that weathermen were routinely the butts of jokes and ridicule.

ALL AT SEA

As we have noted already, detailed weather records date back only to the 1870s in some places, and not even that far in most others. As recently as 1940, weather data for about 90 percent of the planet—including most of the oceans—remained patchy or nonexistent.

The weather at sea presented the greatest forecasting challenges. While meteorological conditions in shipping lanes would regularly be recorded in ships' logs, outside those swaths lay large information vacuums. Even within the shipping corridors themselves, there would be little documentation of major storms. In one of the ironies of the history of weather research, the more timely the information about a marine gale or tropical cyclone, the less forthcoming would be future data about that same storm.

The scenario that repeated itself over and over went something like this: A ship encountered a tropical storm at sea, its weather officer recorded the winds and barometric pressure, and the captain radioed that data to shore. Another ship, at a slightly different location, did the same. Of course, the two sets of wind data were different because tropical cyclones are circular, with the wind running counterclockwise (in the northern hemisphere). Meteorologists at the receiving station could therefore triangulate this incoming information and determine the position of the center of the storm, as well as the speed and direction it was drifting. So far, so good. The shore station then radioed a storm warning to all ships in that general vicinity. Again, so far so good.

Naturally, what happened next was that all shipping made as much speed as possible out of the storm region. As a result, no further observations were made there. Over the following days, the storm might escalate, subside, stall, or change direction; but nobody would know for sure, because there was nobody in the neighborhood to make any observations.

Furthermore, it was not uncommon for meteorologists to "lose" a tropical cyclone in this manner, only to have it strike someplace where it was not expected. Such periods of data scarcity were always tense ones for the forecasters. Grady Norton, the first director of the Miami Hurricane Center (later renamed the National Hurricane Center) described his own approach as follows:

> Whenever I have a difficult challenge in deciding and planning where and when to issue hurricane warnings, I usually stroll out of the office onto the roof, put my foot on the parapet ledge, look out over the Everglades, and say a little prayer."

A NEW SCIENCE

This problem of "lost" hurricanes would not be solved until after about 1970, with the advent of continuous monitoring of global weather from satellites.

It was not that the earlier atmospheric scientists didn't try. They launched balloons with weather instruments. They dropped instruments from aircraft. After the mid-1940s, they dispatched planes to fly actually into the centers of tropical storms and hurricanes. By the 1950s, they were adapting military radar to track storms—at least those that were not too far beyond the horizon. A complete picture of global weather on any given day, however, was possible only after a network of satellites was reliably collecting the data. And because an accurate understanding of climate required an extensive database of historical weather, climatology as a science progressed rather slowly.

Although the U.S. Weather Bureau (renamed the National Weather Service in 1970) did establish an Office of Climatology around 1957, its original function was to archive masses of historical weather data, rather than to conduct any studies of its own. That office evolved into the National Climatic Data Center. Then, in the 1980s, the Climate Analysis Center was established, ultimately to be renamed the Climate Prediction Center. So it was not so very long ago that climatology, after progressing rather slowly, actually emerged as a modern physical science.

METEOROLOGICAL MATHEMATICS

Climate is a monstrously complex subject, involving dozens of variables that interact in multiple ways over long periods of time.

Meteorology, which is more immediate, is only slightly less difficult. Yet why are these subjects so challenging? In 1939, the Canadian meteorologist George Simpson set out the issue like this:

> Meteorology is a branch of physics, and physics makes use of two powerful tools: experiment and mathematics. The first of these tools is denied to the meteorologist and the second does not prove of much use to him in climatological problems.

That statement came at a time when there were no digital computers, the only calculating devices being mechanical adding machines or the specialized analog devices found in an aircraft's bombsite, in planetariums, and in navigation equipment. None of those contraptions were of any use in studying climate. If you wanted to calculate, and if it involved more than adding and subtracting, then you had to perform all the calculations by hand.

Simpson was, however, only partially correct. Scientists *did* try to calculate climatic variables to the extent they could. As far back as the 1820s, the mathematician Joseph Fourier had computed the equilibrium temperature of Earth. He did so under the assumption that the incoming solar radiation was balanced by the planet's outgoing thermal radiation. His results gave him an average temperature that was well below freezing. Since this answer was obviously wrong, Fourier concluded that his underlying assumption was incorrect. Possibly, Earth's atmosphere was trapping some of the outgoing heat. However, he could think of no way to determine how much.

By the 1890s, improved data on Earth's radiation balance became available. In 1896, Arrhenius repeated Fourier's computation—but instead of treating the planet as a single sphere, he carried out laborious, separate calculations for each band of latitude. When he combined all the results, he arrived at an average global temperature of about 0°F (-18°C)—which was about what Fourier had calculated. Arrhenius, like Fourier, concluded that a radiation-balance model was an oversimplification. Clearly, Earth's atmosphere plays a major role in determining its temperature.

THE GREENHOUSE MISNOMER

If the atmosphere were fully transparent to radiation in both directions—incoming and outgoing—it could not possibly have any effect on Earth's average temperature. And indeed, nitrogen and oxygen—the atmosphere's two main constituents—are transparent to both visible and infrared light. The only atmospheric molecules that could possibly trap outgoing heat were CO_2, water vapor, and methane. As we saw back in Table II, these compounds only exist in the atmosphere in meager concentrations.

From the earliest days, the atmospheric heat-trapping process was referred to as the "greenhouse effect," and the gases responsible for it were called "greenhouse gases." These misnomers are still widely used today. Actually, the atmosphere traps heat through a process quite different from the glass of a greenhouse. A glass roof prevents warm air from expanding and rising; in other words, it prevents convective heat loss. The greenhouse gases in the atmosphere do no such thing.

What they actually do is absorb outgoing thermal (infrared) radiation, and then they transfer this heat—through molecular collisions—to the more plentiful nitrogen and oxygen molecules in the atmosphere. This warmed air also radiates heat, but unlike Earth's surface (which radiates heat only upward), each layer of atmosphere radiates in all directions—up, down, and sideways. Some of this re-radiation is therefore absorbed by other layers of the atmosphere—above, below, and even by the ground. Only the thermal radiation from the upper atmosphere (above about 6 miles [10 km]) escapes appreciably into space. In fact, this is the layer where temperature averages roughly 0°F (-18°C), pretty much as Fourier and Arrhenius had calculated.

Yet if the so-called greenhouse gases explained Earth's current temperate climate, then what could possibly explain the ice ages? This was the great climatological puzzle of the 1930s, and many ideas were proposed. Some hypotheses contained a kernel of truth, while others were half-cocked at best. Eventually, the story that emerged went something along the following lines: Water vapor and CO_2 molecules do not stay in the atmosphere forever; they flow in and out in a continuous cycle. The influx part is easy: Water vapor comes from evaporation and from volcanoes; CO_2 comes from decaying vegetation, fires, volcanoes, and as outgassing from the seas. So where do these molecules go?

We know that CO_2 is taken up by plants, by some types of sea life, by being dissolved in rainwater, through the formation of certain minerals, etc.—and a tiny portion even freezes out of the atmosphere (during winters near the poles). Water vapor, of course, also condenses and/or freezes and becomes precipitation from the atmosphere as rain or snow.

IMBALANCE EFFECTS

However, suppose there comes a period when all of these processes become out of balance, when more greenhouse gases are being sequestered than are being released. Maybe there could be a few decades with no volcanic eruptions, or maybe a huge algae bloom in the tropical oceans. With less greenhouse gas in the atmosphere, the planet would cool slightly and so more precipitation would fall as snow rather than rain.

All of this would increase the planet's albedo, so that more sunlight would be reflected and less absorbed, leading to even greater cooling. The polar ice caps expand, continental ice sheets grow. Meanwhile, the cooler seas absorb increasing amounts of CO_2 from the atmosphere. The combined process is driven by "positive feedback," in which all of the changes reinforce one another. Before long, as a result, the planet becomes a much colder place. Eventually, however, the temperature drop does stabilize, because a colder Earth loses less thermal radiation into space.

COMPUTERS AND COW GAS

This hypothesis for explaining the ice ages, however, could not be tested experimentally, because there's no way to run an experiment on a whole planet. Nor could it be calculated math-

ematically when first proposed in the 1930s. There were too many equations with coupled variables to solve by hand. Then, in the 1950s, digital computers became available. Correctly programmed, they were capable of performing calculations considerably more complex than any human alone could tackle.

Those earliest computer results were disappointingly inconsistent. Some predicted an imminent return to a deep ice age. Others predicted a runaway rise in temperatures. None of them predicted a stable climate.

Nevertheless, a great deal was learned from those early computational failures. Atmospheric circulation, for instance, could not be ignored, nor could ocean currents such as the Gulf Stream. They also showed that there needed to be better data relating to some of the variables—such as the "parasol effect" of volcanic eruptions, the albedos of clouds of different types, deep-sea temperatures (which affect the sequestering of CO_2), and even the amount of methane collectively belched by the world's cattle.

Although such data was essential to the development of valid climate models, the American press had a field day ridiculing some of those studies: "Cow gas? You mean that somebody got government funding to study *that*?"

CHAPTER 8

Computers, Chaos, and
Strange Attractors

Increased confusion is valuable when it pushes scientists to get a better answer.
SPENCER WEART

As of 1962, nobody had developed a computer model that could show why the current world climate is distributed as it is, let alone give a clear picture of where the future global climate is headed. By one count, there were at least 54 distinct unevaluated hypotheses about climate change floating around at that time. Six years later, when the number of plausible hypotheses had increased to at least 60, the computer models were still no more definitive. Scientists of the time were justifiably skeptical of the predictions of everyone's computer simulations, including their own.

PROBLEMATIC MODELS

If you examine a world map, you find that London is farther north than any point in the contiguous United States. Paris lies on about the same latitude as St. John's, Newfoundland; and Rome is as far north as Chicago. So why do those European cities have such temperate climates?

One can say it is because of the effects of the Gulf Stream; but such a result ought to be an answer produced by a good computer model, not be part of the input. Early climate models got other things wrong, too. For example, they would fail to predict the existence of the Sahara Desert; or else they would predict deserts in the wrong places; or they might even make South Africa as cold and icy as the tip of South America. All in all, they were a long way from being useful.

The source of the problem was no mystery. The computers of the 1960s just were not powerful enough. For instance, an IBM System/360 of that vintage had just 64 to 128 kilobytes of memory, yet along with its peripherals it filled a large room. As for graphics, there was not enough memory or speed to even consider that sort of application. All you could do with these machines was crunch some numbers. Even then, this function was only possible within limits—unless you wanted to wait days for the results of a single set of computations.

BUDYKO AND LORENZ

In spite of those computational limits, there were still some provocative findings about climate. A Russian scientist, Mikhail Budyko, was studying some of the Soviet proposals to alter climate by diverting Siberian rivers to flow south rather than north, and by spreading soot on icepacks to decrease their albedos. He and his colleagues developed a series of simplified computer simulations to examine the possible outcomes. To their astonishment, those models predicted more than one stable state of glaciation, depending on whether the final equilibrium temperature was reached via warming or cooling. Furthermore, the models predicted several "critical points" (since then, often referred to as "tipping points"). These points are akin to balancing an egg on the peak of a gable roof, where the faintest whisper of a wind will send it careening in either one direction or in the exact opposite direction.

Budyko's conclusion was that climate stability is probably just an illusion. He advised the Soviet government against the proposed climate modification schemes.

At about the same time, 1962, the American meteorologist Edward Lorenz was running his own computer simulations of moving weather patterns, which involved solving 12 simultaneous differential equations. When he terminated one of his snail-paced computer runs, a key variable had the value 0.506127 (it was a dimensionless ratio with no units). Later, Lorenz decided to extend that particular simulation further into the future. Rather than starting all over at the beginning, he resumed it by entering the truncated value 0.506, since he figured that the last three digits were too small to be measurable anyway.

Later, when he was about to document his results, it occurred to Lorenz that it would be sloppy science to report a computer run that had been stopped midway and restarted later. So, for archival purposes, he decided to repeat the complete run from beginning to end. The new results astonished him. After the truncation point, the two runs increasingly diverged. That small, physically immeasurable 0.000127 difference eventually sent off his computer-simulated weather patterns on two entirely different paths.

After reconfirming this unexpected find and assuring himself it was not due to some weird computer glitch, Lorenz published his results in the *Journal of the Atmospheric Sciences* in 1963. Within months, similar results were being replicated by researchers all around the world. It seemed that slight variations in meteorological variables, too small to be measured, could nevertheless escalate into wildly divergent weather systems. The extrapolation was that if that was the case with weather, then it was probably also true for climate.

Initially, Lorenz referred to this phenomenon as "sensitive dependence on initial conditions." By 1969, he and others were using the term "butterfly effect."

The concept, somewhat updated, is as follows: Suppose a hurricane strikes the Virginia coast and we would like to know how that storm developed. To find out, we examine the archived weather satellite data in reverse. Before it was a hurricane, the tempest was a tropical storm; and before that, a tropical depression. A week or so earlier, it was just a "tropical wave" off the coast of West Africa. Prior to that, it was a wet windy place in the rain forest. Earlier still, instruments could not distinguish it from hundreds of other little gusty spots in the region.

So what caused just *one* of those little pockets of wind to escalate to a full-blown hurricane in Virginia, while the others did no worse than blow away someone's hat? We simply cannot tell, because whatever it was, that agent was too small for our instruments to detect. That 0.000127 difference which Lorenz had truncated might have amounted to no more than a butterfly flitting to the left of a tree rather than to its right. When scientists are confronted with a new and curious idea like this, they often become flippant. Hence, the "butterfly effect," although flying squirrels or fruit bats would have worked just as well as examples.

Lorenz's papers and talks on the subject gave rise to a new field of inquiry, sometimes called "nonlinear dynamics" but more commonly referred to as "chaos theory." In its early years, this discipline blossomed quickly, providing a rich conceptual framework for thinking about a broad range of natural phenomena. These ranged from the growths and crashes of insect populations, to the dynamics of white-water rivers, to the shapes of clouds, to describing how a flag flutters in the wind—and even some unnatural phenomena like the trading movements on the stock market and the fortunes of wars.

THE BRICK WALL OF CHAOS THEORY

Unfortunately, after the initial few decades of striking intellectual insights, chaos theory began to stagnate. In recent years, it has made virtually no progress toward solving its central problem: How does one identify the critical points (or tipping points) of a chaotic system?

Chaos itself is often defined as a dynamic system that has a sensitive dependence on its initial conditions. That, however, borders on tautology—using a concept to explain itself. So a better definition might be:

> A chaotic system is one in which statistics (means, medians, modes, standard deviations, and so on) vary so drastically over time as to not be meaningfully descriptive of the system or its dynamics.

Let me expand on this point. We can compute the positions of the planets for any calendar date, no matter how far into the future, and the result will be predictable within the limits of measurement error. Planetary motions are an example of a deterministic system—one in which the initial conditions completely determine the future. If those initial conditions change slightly, the future also changes slightly.

On the other hand, certain other phenomena are only *statistically* determinate. As an example, we cannot predict how long we personally will live, yet we can consult actuarial tables to find the probability of our living to a certain age, given that we were born in a certain year. Such statistics and probabilities do not predict the futures of individuals, but rather large populations of individuals (which, in fact, is what makes the insurance industry viable). Statistics, of course, may also change over time. In fact, that is exactly what's been happening with human life expectancies. If you were born in 1980, you're likely to have a longer lifespan than most people born in 1940. (Not only that, but your feet are also likely to be bigger, on average, than those of earlier generations.) Again, though, small variations in initial statistics do not result in wildly varying future statistics. Deterministic systems as well as statistically deterministic systems tend to be reasonably well behaved.

There are, however, phenomenological systems where statistical measures vary wildly from one time interval to another. Count the number of tropical storms striking North American coasts, for instance, and calculate the 10-year average. Does this say anything about the average in the previous or the following decades? No, it turns out, it does not. The number, intensity, and timing of tropical storms in one decade are poor predictors of their number, intensity, or timing in any other decade. In other words, hurricane data is *chaotic*.

According to chaos theory, this unruliness results when a complex system (one with many interacting variables) involves something called a "nonlinear" process. Consider wind, for instance. A 10-mph wind blowing over the sea reduces the barometric pressure slightly—by about 0.005 inches (0.013 cm) of mercury. Double the wind speed, however, and the decrease does not double; the barometer actually dips by about 0.02

inches (0.05 cm). But increase the speed by a factor of ten (all other factors being equal) and the pressure drop due to the wind alone is now almost 0.5 inch (1.3 cm). This is what we mean by a nonlinear phenomenon. A change in one quantity does *not* affect the other quantity proportionally.

Moreover, this interaction goes both ways. It is not just wind that affects barometric pressure, but pressure differences also drive the wind. These two variables, therefore, interact in a complex feedback loop—and both are affected by other variables such as air temperature, water temperature, and humidity.

Consequently, as a system, the atmosphere exhibits an extreme dependence on very minor fluctuations in the variables—fluctuations which, in fact, may sometimes be too small to be measured. Two tropical storms that appear initially to be pretty much the same may exhibit divergent behaviors. For instance, one may escalate and strike shore as a major hurricane, while the other changes direction, subsides, and dies out harmlessly at sea.

STRANGE ATTRACTORS

Given that statistics do not apply to chaotic systems over the long term, what insights could chaos theory possibly offer? The main one is the concept of the "strange attractor."

Consider a run of rapids on a river. In fact, consider a particular river—the Niagara—a few miles downstream from the famous falls. Suppose you conduct a little experiment—say, tossing a few dozen numbered tennis balls into the froth. Farther downstream, a colleague of yours records the time it takes each ball to reach him or her. Some balls arrive there within minutes. Others, however, take an hour or more. A few fail to show up until a day later, because they were temporarily trapped in an eddy somewhere. Although the distance divided by the average time gives you a statistical average speed, such a result tells you virtually nothing about the actual behavior of the river. Nor does it tell you anything about what happens to the next ball you might toss in.

Several miles downstream of the falls, the Niagara River makes a 90-degree turn. On that bend is a huge whirlpool. When the river is high, the water circulates in a counterclockwise direction; but when it is low, the circulation is clockwise. In the terminology of chaos theory, the whirlpool has two "strange attractors": one corresponding to the set of dynamical variables at high water, the other corresponding to low water. Although the statistical travel times of our tennis balls in this turbulence remain essentially meaningless, those two whirlpools are nevertheless each quite real and quite stable.

If you watch the high-water whirlpool while the river is dropping, very little seems to change—until a critical point is reached. Then, suddenly, the river begins to appear confused. If the water level then continues to drop, within a half hour or so the vortex reestablishes itself in the opposite direction. In other words, when that critical point (or tipping point) is crossed, the river changes its behavior fairly quickly.

Just like the Niagara whirlpool, climate change defies conventional long-term statistical analysis because:

a) the underlying phenomena are nonlinear; and
b) climate has the ability to "flip" relatively quickly from one dynamical state to a completely different state.

If you average one state or set of conditions with another, the result is nonsense. If you average what is happening near a tipping point, you may also get nonsense.

Earth's climate seems to have at least three strange attractors. One (maybe more) corresponds to the moderate climatic conditions we currently experience, and that have been around for roughly the past 10,000 years. Another corresponds to a sweltering Earth with a runaway greenhouse effect. A third is a "white Earth" strange attractor—another ice age. Furthermore, according to Mikhail Budyko and others, there may be more than one ice-age attractor, characterized by different ambient global temperatures and different degrees of glaciation.

Every modern climatologist is fairly certain that such strange attractors are real enough. Moreover, if they are real, then there are also real tipping points able to flip the climate from one strange attractor to another. The climatological challenge is to determine just where those tipping points are. Unfortunately, this is turning out to be a monstrously difficult task.

QUESTIONS OF BALANCE

Now that I've said a few things about chaos theory, I need to contradict myself. Not by much, but a little. My clarification has to do with the usefulness of statistical averages in a chaotic system.

In the tennis ball example, I said that the average travel time through the Niagara rapids is a meaningless quantity. That still stands. Knowing that number gives you absolutely no information about the behavior of the next ball you might toss into the froth.

Yet there *is* one meaningful statistical quantity in this system, and that is the average flow rate of the river. You can't measure it in the rapids themselves, but you *can* do so upstream or downstream, where the flow is more well-behaved. On average, 100,000 cubic feet of water (2.8 million liters) enters the rapids per second, and on average this flow is the same leaving the lower end. Obviously, then, this is also the average flow within the rapids themselves—even though it cannot actually be measured there.

Now, suppose that there's a point in time when the entering flow increases to 130,000 cubic feet per second (ft^3/s) and the exiting flow is still 100,000 ft^3/s. The inference then is that the water in the rapids is rising, because what's going in is not balanced by what's going out. Notice that to reach this conclusion, we don't need to measure the depth of the river, nor the heights of the waves, nor any of the other details. If more water is going in than is coming out, then the water in between must be getting higher.

Similarly, if we look at our planet's radiation balance, and if what is coming in from the Sun is not balanced by what is going out into space, then we can conclude that the planet is heating up (on average). To reach this conclusion, we do not need to know the actual average global temperature—which remains, essentially, a meaningless quantity.

MONSTER MATH

Not surprisingly, computerized climate models have improved significantly since the 1960s. Not only is the computational technology vastly better, but so are the available data. On a daily basis, we are now able to monitor the whole globe and record an array of surface temperatures, cloud cover, sea levels, snow cover, the movements of storm systems, and so on.

There remain major challenges, however. Most of today's climate models rely on an imaginary grid blanketing Earth's surface, each section typically measuring one degree of longitude by one degree of latitude (which averages roughly 69 miles on a side). On top of these distorted squares, the modelers usually stack five or six layers of atmosphere—resulting in about 700,000 atmospheric boxes. These cubicles aren't exactly cubes, however; they're rounded because of the Earth's curvature, and they're flatter than they are wide. Some models also include a few layers that extend down into the oceans.

Each of these 700,000 or more cubicles is then described by a set of atmospheric parameters: temperature, barometric pressure, humidity, albedo, greenhouse gas content, and so on. On average, each cubicle borders six other cubicles, so there are over 4 million faces over which something—heat, wind, clouds, and so on—can flow. There are at least a dozen equations (most of them what are called "second-order partial differential equations") that describe these "transports." So approximately 56 million equations need to be solved in order to carry the model from one instant in time to the next point in time—say, an hour later, or maybe a day later. Thus, if the model is clocked hourly, it takes 480 billion rather complex calculations to simulate one year of weather. Projecting forward 20 years requires about 10 trillion calculations. Remember, too, that these calculations are not simply adding and subtracting (most of the processes, after all, are nonlinear). Furthermore, each step in the computation involves the simultaneous solution of at least a dozen coupled equations. This is very serious, high-powered mathematics.

Now it might seem that an all-guns, brute-force approach like this should tell us everything we might want to know about Earth's future climate. Unfortunately, it does not. For one thing, a "one-size-fits-all" grid does not conform very well to the problem. Earth's climatic regions, as pointed out in Chapter 3, come in a vast range of sizes and shapes.

So why not just change the grid so that it conforms to the shapes of the climatic regions? Alas, this cannot be done, because those regions are shifting, and their future shifts are one of the things we want to know about. Instead, what needs to be done —in fact, is being done—is to incorporate increasing detail into climate models in the most critical geographical regions.

There is also the problem of the availability of data. Despite the incredible number of measurements being made continuously from satellites, there remain some gaps. For instance, satellite instrumentation cannot see all that well through cloud cover, particularly when one layer of clouds drifts over another at a lower altitude. Nor can orbiting instruments see into water—and remember, the seas cover about 70 percent of Earth's surface. Although these problems are slowly being remedied, it will still

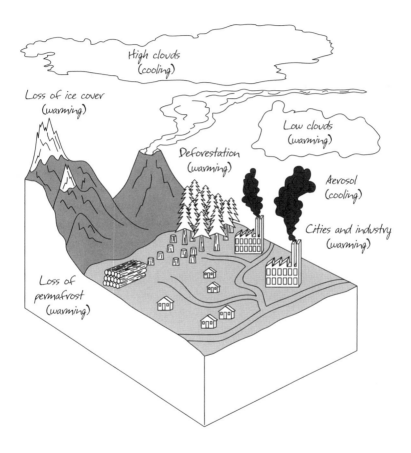

FIGURE 7: CLIMATE WARMING AND COOLING

Climatological models incorporate both warming and cooling processes. All of the current models reveal an imbalance in favor of warming. If human activities don't change, this heating imbalance will continue.

take many years before a comprehensive data set of the planet's climate is archived.

So what do the current climate simulations tell us? The answer is: quite a lot, even though some of the numbers remain imprecise.

- Earth is currently absorbing more sunlight than it is radiating back into space—an energy imbalance of about one part per thousand that is global and is increasing.
- A rise in atmospheric CO_2, and to a lesser extent methane, is causing the energy imbalance.
- Human activity seems to be responsible for most of the increase in greenhouse gases, as there are no purely natural processes that can account for more than a small fraction of the increase.
- If all human generation of greenhouse gases is halted today, there will still be a time lag of several decades to as much as 50 years before the average global temperature stabilizes.
- Long before then, the Arctic Ocean will be essentially ice-free in the summer.
- At first, global agricultural production will increase. After an average rise of a few degrees, it will decrease.
- The distribution of deserts and grasslands will shift, but in ways that are not fully predictable (at least not currently). There is a possibility that the Great Plains could become a huge desert.
- In Africa, the northern, sparsely populated regions will

probably have more rainfall, while the southern, more populated areas will get less.

- Some plant and animal life will expand into newly created environmental niches. (Mosquitoes and other insects, for instance, will thrive at higher altitudes; and insect-borne diseases such as malaria will thus gain new footholds.)
- Other species will not be able to migrate or adapt quickly enough to their changing habitats and will become extinct.
- The seas will become increasingly acidic, killing coral and affecting marine animals, with the entire marine food chain altered drastically.
- There are likely to be more tropical storms and hurricanes—chaos theory notwithstanding. (The jury is still out on whether or not those storms will be more intense, but that is a definite possibility.)
- Tropical storms will hammer coastlines that presently are not at risk from them or which get them rarely (Brazil and New England, for instance), and there may even be hurricanes in the Mediterranean Sea.
- Heat waves and droughts will be increasingly common in some regions, increasing health risks and the prospect of more frequent forest fires.
- Rising sea levels will threaten numerous coastal cities and towns, with some inhabited Pacific atolls virtually disappearing.
- The Gulf Stream (as well as other parts of the so-called Meridional Overturning Circulation) will stutter and slow down due to dilution by meltwater from the Greenland

ice sheet. It may even stall completely, plunging northern Europe into a deep freeze as other parts of the globe warm up.

- Thawing of the permafrost in Alaska, northern Canada, and northern Russia will lead to structural instabilities, the shrinkage or disappearance of thousands of lakes, and disruptions to fish and game populations.
- Other places will experience violent floods in the late winter and early spring, as highland precipitation that normally falls as snow will instead fall as rain and run off immediately.
- There is a tipping point, beyond which it will be impossible for humans to do anything to stop these and related changes. Some climate simulations find this tipping point at an increase of about 4.9°F (2.7°C). At least one has suggested that a mere 2.2°F (1.2°C) will do it. There is little doubt that a change of 5.4°F (3°C) is beyond the tipping point, because at that average temperature the planet's natural carbon cycle will effectively reverse, with microbial action in the forest beds releasing huge quantities of additional CO_2 into the atmosphere and driving further, quite drastic, climate changes.

Now all of the above predictions can be disparaged on the basis that computational projections are not reality. On the other hand, most of the past simulation errors have been in the direction of underestimating the threat, not overestimating it. In particular the following phenomena were underestimated:

- The Arctic sea ice has been melting earlier than most of the earlier models projected;
- atmospheric CO_2 concentrations have been rising faster than predicted; and
- average global temperatures have risen faster than most models have forecast.

IGNORING THE SCIENCE

If this is not enough to worry about, there's also another threat: certain influential leaders who steadfastly ignore the science. As recently as July 30, 2009, for instance, the former House Majority Leader Dick Armey (R-Texas) in the U.S. Congress made the following statement at a bicameral hearing on proposed climate legislation:

> Let me say I take it as an article of faith if the Lord God Almighty made the heavens and Earth, and he made them to his satisfaction, it is quite pretentious of we little weaklings here on Earth to think that, that we are going to destroy God's creation.

As we have already seen, the science suggests otherwise. It is, in fact, quite possible for us "little weaklings" to destroy our environment as we know it—regardless of whether or not it was created by a deity.

CHAPTER 9

The Buck Stops Where?

We have met the enemy and he is us.
COMIC STRIP CHARACTER POGO, CREATED BY WALT KELLY

To be alive is to affect the environment. In fact, it is unlikely that there was ever a time when a human community did not acknowledge this in some way or another, albeit in the context of their time and culture. An ancient shepherd allows his flock to overgraze a plot of land and the soil soon erodes. A settlement neglects to maintain its latrine and people start getting diseases. A tribe eats too many pigeon eggs and the pigeons disappear.

THE PEOPLE EFFECT

Back in antiquity, the most obvious of such problems had simple fixes—for instance, relocate the whole community or maybe break it up and disperse from time to time. But with the coming of the Industrial Revolution in the early 1800s came more insidious forms of pollution—ones that were not so easy to remedy. Smelters befouled the air with soot and other pollutants, acids and heavy metals leaching from mine tailings poisoned streams and lakes, and it was impossible to cross a street without tramping through horse manure that teemed with bacteria.

In addition, the increased populations that were attracted (and required) by the new industries aggravated the human toll by clustering close to their source of employment. There was no such thing as wastewater treatment, so every stream running through any town became an open sewer. Technology to address such problems simply did not yet exist, for such developments always lag behind the forces that drive them—sometimes by a considerable length of time.

It never occurred to anyone in the 1800s that pollution was more than a local issue. If you traveled just a few miles away from the cities and the mines, there was never any problem in finding clean air and pristine water. At the time, any concept that human activity might someday affect the *global* environment was too farfetched to merit consideration.

A large number of people have been added to the planet since those early days. Table IV shows how the world population has grown since the end of the last ice age. The one million or so humans alive in 10,000 BCE could not have possibly impacted the global environment. By 1800, our human numbers had climbed by a factor of a thousand to just under a billion. At that time, human-generated CO_2 was barely beginning to escalate.

Table IV: World population growth

Date	World Population
10,000 BCE	~1 million
2000 BCE	~35 million
1000 BCE	~50 million
500 BCE	~100 million
1	~200 million
1000	~310 million
1750	791 million
1800	978 million
1850	1.2 billion
1950	2.52 billion
1960	2.98 billion
1970	3.69 billion
1980	4.43 billion
1990	5.26 billion
2000	6.07 billion
2008	6.71 billion

Today, there are seven people alive for every one person who was breathing in 1800. Moreover, although many today still subsist at poverty levels, many others do not. Those billions of us lucky enough to live comfortably have a tremendous collective impact on Earth's atmosphere and hydrosphere. And that impact is growing because of our increasing prosperity *and* our increasing numbers.

Consider this: If you were to spread out the world's population evenly on all of our planet's landmasses, excluding Antarctica, it would mean 112 people on every square mile—including deserts and mountaintops. Nowhere on Earth's surface would you be able to walk more than a block (about 530 feet [160 m]) in any direction without encountering another human being. Additionally, almost every person you encountered would be generating greenhouse gases through his/her daily activities: driving, cooking, manufacturing, consuming electricity, and so on. It would be remarkable if all of that bustle *did not* affect both the atmosphere and, ultimately, the seas.

EARLY WARNING

The first wake-up call about anthropogenic global environmental effects came in 1962 with Rachel Carson's book *Silent Spring.* In this groundbreaking publication, she developed a carefully reasoned case that pesticides and insecticides were being overused and were also inadvertently exterminating wildlife—particularly birds—around the world.

Before publication, her book was reviewed for accuracy and credibility by a dozen or so other scientists. Although Carson was careful to state that she was in no way campaigning against the use of dichlorodiphenyltrichloroethane (DDT) to control mosquito-borne diseases, her disclaimers fell on some deaf ears. The agricultural chemical industry—particularly Monsanto, Velsicol, and American Cyanamid—mounted a vicious campaign against her, claiming that she was unqualified (although she was a graduated biologist with numerous articles and three science books already to her credit), and threatening her with lawsuits for libel. However, Carson never actually had to appear in court, because she died of cancer in 1964.

Most science historians credit Rachel Carson with spawning the environmentalist movement, as well as setting the wheels in motion that led to the creation of the U.S. Environmental Protection Agency (EPA) in 1970 under President Richard Nixon. Yet, beyond that, it was Carson's work that led scientists to realize that we humans are not a passive part of the global environment; our collective activities are, in fact, capable of changing our planet.

In the decades since Rachel Carson's death, the world population has more than doubled. That makes one of her statements doubly ominous:

> Only within the moment of time represented by the present century has one species—man—acquired significant power to alter the nature of his world.

For better—and, yes, for worse.

Following Rachel Carson's lead, scientists began to link other international environmental effects to human activities. Two of those discoveries are worth a brief mention.

ACID RAIN

The acid rain issue surfaced in Europe during the industrial expansion of the 1960s, when whole forests downwind of new smokestacks began to wither and die. By the 1980s, this phenomenon was well documented, as even lakes were becoming acidic, killing the fish and frogs. Furthermore, this was not just happening in Europe, but also in many parts of the United States and Canada that were far from industrial activity.

The culprit was easy to identify: the gaseous oxides of sulfur and nitrogen generated during fossil-fuel combustion combined chemically with atmospheric moisture to create sulfuric and nitric acids. Not only does the resulting acid rain harm flora and fauna, but it also dissolves brick and stone, including in some cases buildings and statuary dating from antiquity.

Although the acid rain phenomenon was clearly international—with one nation's pollution raining down on another—it was not quite global in scope. In most cases, bilateral treaties addressed the problem successfully by limiting and/or reducing the offending smokestack pollutants.

If DDT and acid rain raised awareness that environmental pollution may have international implications, the issue of ozone depletion went many leagues beyond that. Here, for the first time, the issue was unambiguously global in scope.

HOLE IN THE OZONE

Although the form (allotrope) of oxygen called ozone (O_3) contributes to smog when it is close to the ground, in the stratosphere it serves to shield us from the Sun's cancer-inducing ultraviolet rays. In the mid-1970s, instrumentation on weather balloons revealed that Earth's ozone layer was thinning. In Antarctica, where by 1985 the ozone concentration had declined by as much as two-thirds, this depletion progressed to a so-called "hole."

By then, the culprit had already been identified as chlorofluorocarbons (CFCs, or freons). As gases, they were used as refrigerants and as propellants in spray cans, and were diffusing throughout Earth's atmosphere. At high altitudes the Sun's ultraviolet light was liberating the chlorine atoms from the CFCs. Chlorine, it turns out, is particularly extreme in its effect on ozone: A single chlorine atom is capable of catalyzing the destruction of several thousand ozone molecules before it either rains back down or drifts off into space.

This revelation was a shocker. There is not a great deal of freon on the planet, and 100 percent of it is anthropogenic. The only way it could occur in the atmosphere was from spray cans or by leaking from refrigerators and air conditioners. "Common sense" told everyone that it was impossible for such small amounts to have detrimental global effects. Yet, as so often happens in science, common sense was wrong. No alternative explanation was possible: CFCs were destroying Earth's natural ultraviolet shield.

The solution was an international ban on the use of CFCs. In 1985, the Vienna Convention for the Protection of the Ozone Layer was signed by 20 nations. That event laid the groundwork for the 1987 Montreal Protocol, an international treaty that set timelines for the reduction and ultimate ban of ozone-depleting chemicals.

Amazingly, the treaty has worked. Earth's ozone layer has shown clear signs of recovery. Equally surprising is the fact that there are no complaints from the chemical industry, which is happily producing and marketing alternative refrigerants—which, of course, are priced higher than the original chlorofluorocarbons they replaced.

The IPCC

In 1988, James Hansen, director of the Goddard Institute for Space Studies (GISS) testified before the U.S. Senate as to the prospect of anthropogenic global climate change. Hansen is hardly a wild-eyed environmental activist; his political views lean toward the conservative, and he had never before mixed his science with politics. However, his research team had carried out a series of climate simulations using three alternative scenarios relating to future greenhouse gas emissions. The results were scary. In one case, where they included a major volcanic eruption, the so-called "parasol effect" did indeed cool the planet slightly for a couple of years. Ultimately, however, in all three scenarios, Earth grew warmer.

Hansen knew full well that he had no personal monopoly on scientific truth. He combed the scientific literature to see what other researchers around the world were finding. Some had concluded that the warming would be more modest than his predictions, while others predicted greater degrees of warming. However, not a single scientific study published anywhere predicted global temperature stability, let alone cooling.

In his U.S. Senate testimony, Hansen stated that he was "99 percent sure" that an anthropogenic greenhouse effect had been detected and that it was already changing Earth's climate. He followed with this statement to a group of reporters:

> It is time to stop waffling . . . and say that the green-
> house effect is here and is affecting our climate now.

That same year, 1988, the United Nations established its Intergovernmental Panel on Climate Change (IPCC), tasking its members to evaluate the risk of climate change caused by human activity. The 60 or so panelists, all scientists, were appointed by the World Meteorological Organization (WMO) and the United Nations Environment Programme (UNEP).

Rather than conducting its own research directly, the IPCC reviews and evaluates the complete spectrum of climate-related research being conducted by others around the globe. If there were any credible scientific studies *anywhere* that contradicted the climate change hypothesis, this is where they would show up.

No such contradictory results ever surfaced. Yes, there were numerous articles where nonscientists ventured personal opinions, unsupported by any scientific evidence, claiming that climate change was a hoax. There was also a smattering of company-sponsored research where the investigators cherry-picked data supporting that particular company's economic interests (e.g., the continued burning of fossil fuels). But as for unbiased research, there was absolutely none that offered any evidence that global climate was *not* changing.

Nevertheless, the panel was cautious about its conclusions. In its first official report, issued in 1990, the IPCC stated that the unequivocal detection of a human-enhanced greenhouse effect was not likely for another decade or more. On the other hand, the 60 authors also stated that the *probability* was very high of an anthropogenic greenhouse effect being underway—and with human-generated CO_2 being responsible for more than half of it. Further, they predicted that the average global temperature is likely to rise by 0.5°F (0.3°C) per decade over the following century.

In 1992, in anticipation of a United Nations–sponsored Earth Summit in Rio de Janeiro, the IPCC issued a supplementary report in which it reconfirmed its previous conclusions and called for additional climate research. That summit resulted in an international treaty—the United Nations Framework Convention on Climate Change. With 192 signatory nations, including the United States and Australia, the Convention's stated objective was:

> To achieve stabilization of greenhouse gas concentrations in the atmosphere at a level that would prevent dangerous anthropogenic interference with the climate system.

Because the climatological data was still somewhat imprecise, however, the treaty did not set any mandatory limits on greenhouse gas emissions for individual nations, nor did it contain any enforcement provisions. Instead, it provided for future updates (or "protocols") to address those matters. That treaty, usually referred to as the UNFCCC, came into force on March 21, 1994.

Meanwhile, the IPCC continued its work, issuing its second

complete report in 1995. That document concluded that every new climatological study in the past five years had confirmed that the problem was clearly worsening—and that, yes, human activity was responsible. The panel also pointed out, however, that there remained uncertainties about the precise numbers, and it suggested specific areas of further investigation that would help.

THE KYOTO PROTOCOL

In December of 1997, signatories to the UNFCCC met in Kyoto, Japan, to work out a protocol (now referred to as the "Kyoto Protocol") to reduce anthropogenic greenhouse gases on a global scale. The result was a complicated formula involving different emission caps for different "developed" countries and globally traded "carbon credits" to assist the nations in meeting those caps. The target was modest: an average reduction of 5.2 percent from 1990 levels by the year 2012. The protocol went into effect in early 2005 and will expire in 2013. Conspicuously absent from the 182 parties ratifying this agreement were the United States and Australia—although Australia ratified it a few years later. As of this writing, the United States has not; and so one can only be hopeful that it will be a party to the international treaty on climate change that supersedes Kyoto after 2013.

In 2001, while the Kyoto Protocol was still being ratified by most of its signatory nations, the IPCC issued its third assessment report. This time, some of the more tentative language of the earlier reports disappeared. None of the prior conclusions were revoked. In fact, all had been reinforced by investigations that explored at least 35 alternative scenarios for the future of the planet's climate.

The report stated that, under every conceivable set of assumptions about the hazier parameters, human influences would still continue to change the atmospheric composition throughout the twenty-first century. By 2100, the average global temperature will rise by 2.5°–10.4°F (1.4°–5.8°C). During the same period, sea level will rise by 4 inches to 3 feet (10–90 cm). While it is true that these are large ranges of uncertainty, the projected numbers were never zero, nor did they ever appear as negative.

The most recent IPCC report was issued in 2007. Warming of the climate system is now described as being "unequivocal." Even if atmospheric greenhouse gas concentrations were to be stabilized immediately, anthropogenic warming and sea-level rise will continue for the next several centuries.

Just as unequivocal is the use of "anthropogenic" here. The probability that these trends are due to natural climatic variations is less than 5 percent. There is greater than a 90 percent probability that we will experience more frequent heat waves and heavy rains. Coupled with this is the 66+ percent chance that some regions will suffer an increase in droughts, tropical cyclones, and catastrophic storm tides. Atmospheric concentrations of CO_2, methane, and nitrous oxide are now higher than at any time during the past 650,000 years. Worse, even with any (impossible) immediate cessation of all emissions, the current excesses of concentra-

tion will continue to influence Earth's climate for the next 1,000 years.

POLITICAL DIMENSIONS

In the year 2000, the United States held a presidential election in which the two main parties' candidates were the then current Vice President Al Gore and Texas Governor George W. Bush. Gore was a supporter of the Kyoto Protocol, which he had participated in brokering. Bush was against that protocol on the grounds that it might hurt the U.S. economy. The electoral victory went to Bush. In January 2001, Bush was sworn into office.

That left Gore, a lifelong environmentalist, with some time on his hands. When he read the 2001 report of the IPCC, it alarmed him enough to activate him. He put together a slide show about global warming and, over the course of the next five years, he presented it in person at least 1,000 times in various cities and towns across the United States. In 2006, he founded the Alliance for Climate Protection. That same year, he starred in the documentary film *An Inconvenient Truth*, which was based on his traveling slide show, and which won an Academy Award. The following year, Gore and the IPCC were jointly awarded the 2007 Nobel Peace Prize.

While Al Gore was taking climate change seriously, the Bush administration was not. Jokes about the inaction proliferated in the media:

> President Bush has a plan. He says that if we need
> to, we can lower the temperature dramatically just
> by switching from Fahrenheit to Celsius.
>
> — JIMMY KIMMEL

> According to a survey in this week's *Time* maga-
> zine, 85 percent of Americans think global warm-
> ing is happening. The other 15 percent work for the
> White House.
>
> — JAY LENO

And so on ...

A WAR ON SCIENCE

Pervasive in the Bush administration was a remarkable disregard
for science and scientists. Policy decisions were routinely made
on the basis of ideology. Shortly after the election, for instance,
Vice President Dick Cheney assembled an "energy task force" to
guide the administration in developing its energy policies. The
membership of that committee was never made public, but we do
know who *was not* on it: no environmentalists, for instance, and
no scientists. Then, throughout the next eight years, Bush and
Cheney maintained their close ties with the captains of the fossil
fuel industries. To those executives, any talk of global warming
or climate change was anathema; after all, their profits depended
on their customers continuing to convert their products into
atmospheric CO_2.

Political appointees with no scientific background were placed in charge of key U.S. governmental agencies, where they squelched or butchered scientific reports that conflicted with the profit-first ideology. One of those censored was James Hansen at GISS.

Rather than quitting, however (as some other scientists did), Hansen went to the press to complain that the administration was attempting to alter legitimate scientific findings. The Bush administration retaliated by chopping Hansen's research budget. The entire gruesome story would eventually be documented in Mark Bowen's 2007 book *Censoring Science*. Other well-researched books about that period's governmental assault on science included *Undermining Science* by Seth Shulman (2007) and *The Republican War on Science* by Chris Mooney (2005).

This program of intellectual suppression, in other words, was not a figment of someone's imagination; it was pervasive and well documented. In 2005, for instance, the Bush administration actually thanked Exxon executives for their role in determining the nation's policy on climate change and opposing the Kyoto Protocol.

Earlier in this book, I spoke about the power of tenacity as a way of knowing. Repeat a fabrication often enough, and eventually people begin to think of it as a fact. So it was during the Bush years that the U.S. government, the fossil-fuel companies, the Republican Party, and much of the conservative media collaborated to seduce a large segment of the American public into believing that climate change was a hoax, or else that there was still an ongoing scientific debate about the phenomenon. This,

despite the fact that the scientific reality had been well established by the time of the second IPCC report in 1995.

The rest of the world has not been as petulant as the United
States, as evidenced by the 183 other nations that have signed
and ratified the Kyoto Protocol. Just four eligible nations have
not signed—Afghanistan, Brunei, Chad, and San Marino—and a
few others are considered to be "observers" and are not expected
to sign. The United States, as mentioned earlier, has signed but
not ratified.

Yet, until 2008—when it was surpassed by China—the
United States was the world's biggest emitter of CO_2 from power
generation. China, to its credit, had seen this coming, and a year
earlier (June 2007) it unveiled a 62-page plan that placed climate
change at the heart of its energy policies. Meanwhile, however,
China continues to insist that developed countries (China is still
not considered to be one) have an "unshirkable responsibility" to
take the lead on cutting greenhouse gas emissions.

CHAPTER 10

Quo Vadis?

Lead, follow, or get the hell out of the way.
ATTRIBUTED, IN SLIGHTLY DIFFERENT FORMS,
TO GENERAL GEORGE S. PATTON AND THOMAS PAINE

One thing the skeptics have on their side is that global climate change is a slow-motion catastrophe. It is not something you notice on your back porch thermometer in a year—or even a decade—for at least two reasons. First, normal weather variations drown it out over the short term; and, second, your porch is far from being global.

UNDERSTANDING DENIAL

Millions of Americans hear TV pundits like Glenn Beck reporting "facts" such as 1934 being the hottest year on record—which, to Beck, apparently disproves global warming. It matters not that such a statement is simply untrue, nor that the conclusion fails to follow even if the assertion *were* accurate. Millions of us hear tidbits like those from Beck, and many of us nod in agreement as we rise from the couch to get another slice of pizza.

Be that as it may, why do I assert that Beck's claim about 1934 is not true—especially when his data was indeed drawn from GISS scientific reports? Because what GISS *actually* reported was that 1934 was a record hot year *only* for the 48 contiguous states of the United States, which constitute less than two percent of Earth's surface. In those 48 states, the five hottest years of the last century (in order) were indeed 1934, 1998, 1921, 2006, and 1931. In contrast, globally, the five hottest years (again, in order) were 2005, 1998, 2002, 2003, and 2006. Clearly, in the past 100 years, the five hottest were actually within the preceding decade.

It is easy to skip over the significance of the world "global."After all, most of us rarely think globally. (As one relevant blogger expressed it sarcastically, "Huh? You mean the rest of the world exists? It is not all just the United States?")

It is also very easy to misunderstand the science—because so few of us have scientific backgrounds. Although that in no way excuses Fox News or CNN or the *Washington Post* columnists or various politicians for their prevarications, to some degree it may excuse many of the public for becoming so confused about the topic.

Combat Checklist

So how does one make sense of the continuing flurry of rhetoric denying the reality of global climate change? There is no single formula, but the following few guidelines may be helpful.

- Are the purported climatological facts local, regional, or are they actually global? If they're local or regional, then by what method has the commentator extrapolated them to reach a *global* conclusion? Was that method analytical, or was it mere hand waving?
- Are the stated facts single data points or are they whole sets of data taken over a period of time? A record cold day in Timbuktu proves no more than a record hot day in Moscow. It is climate change over the long term that's important.
- Is the data accurate? Backyard thermometers and rain gauges seldom are. How was the data taken and how was it verified?
- Was the data cherry-picked? Some smokers, for instance, live more than 100 years. But does that prove that smoking is not harmful?
- Is the commentator promoting any ideological agenda? Who, for instance, is paying him or her? (After all, everybody is paid by *someone*.)

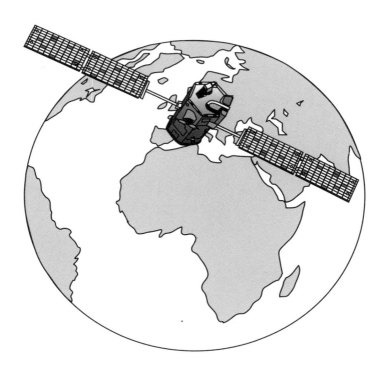

FIGURE 8: GLOBAL TEMPERATURE MEASUREMENT
Variations in global temperature over time are measured by satellites, not by ground stations.

- Is the commentator's claim corroborated by anyone else? If so, by whom? Did any of those corroborators have scientific credentials? If so, who is paying *them*?
- Finally, keep in mind that science is driven by ignorance. Scientists study what they *do not* know, not what they already know. Never be surprised, therefore, if there are new climatological discoveries in the future. This will not mean that the current body of knowledge is wrong, but rather that it is incomplete. In fact, it will always be incomplete.

PROTOCOL PROBLEMS

The Kyoto Protocol was a remarkable human achievement, and some consider that its ratification by virtually all the world's nations was the U.N.'s greatest ever accomplishment. However, "Kyoto" (as it is often abbreviated) does have its flaws as well as more than a few legitimate critics.

Although the baseline year (against which to measure future reductions) was to be 1990, several nations were granted alternative baselines. "Developing" countries (including India and China) were given additional waivers because, at the time of the pre-treaty negotiations, they were not major emitters of greenhouse gases.

China has since become the world's major source of anthropogenic CO_2, but, as noted earlier, the Chinese government has recognized this and is instituting a program to reduce its emissions.

CARBON TRADING

Above all, probably the most controversial aspect of the Kyoto Protocol is its "cap-and-trade" provision. Each participating country has been given a greenhouse gas reduction goal that it has to meet against a set timeline. Any country that beats its goal can then "sell" the surplus of polluting capacity to a nation that is still trying to meet its standards.

As a result, some countries are permitted to continue to emit greenhouse gases as long as they are willing to pay for them; others are free to do so as long as they are still "developing." On top of everything, of course, this all incorporates an assumption that 183 separate national governments are each being honest about their data, and that they are measuring it in the same way.

Within each individual country, the process becomes even more complicated. The allocations need to be parceled out to individual industries, which leads ultimately all the way down to allocating to individual smokestacks. Automatically, things being the way they are, this risks politicizing the system, as well as creating incentives for "creative" accounting. On April 30, 2009, for example, in an editorial in the *Los Angeles Daily Journal*, Robert Benson asked whether the cap-and-trade system might become "the next credit default swaps," in direct allusion to big bank trading that precipitated the global economic meltdown of 2008–2009.

ALTERNATIVES

Is there a simpler way? Yes, but it would require China, the United States, Russia, India, and Japan (currently the biggest greenhouse gas emitters) to take the lead voluntarily. In collaboration, they could agree to reduce their emissions to a level below those of 1990 without trading carbon offsets. Within these major polluting countries, industries might be assessed for a straightforward carbon tax based on their individual emissions. The subsequent revenues could then be applied to the development of future technological improvements that would serve in averting greenhouse gas emissions.

But would not that lead to an economic disaster, just as the Bush administration had so steadfastly maintained? Certainly, it could; but only for certain industries—particularly for those involved with fossil fuels. Many other companies would benefit from the creation of new products for new markets. In light of this, it is very well worth listing a few of the proposed strategies for reducing emissions of greenhouse gases—and considering what might be their general economic impacts.

ENERGY CONSERVATION

A prime strategy is a simple conservation of energy: adding insulation to one's homes and offices, manufacturing appliances that use less electricity, expanding recycling programs, and so on. Such practices can be considered as investments rather than expenditures, insofar as they ultimately save more money than they cost upfront. (While the U.S. Energy Policy Act of 2005 does already include some incentives for conservation, it also encourages the development of new oil fields, exempts oil companies from parts of the Safe Drinking Water Act, and overall is a hodgepodge of concessions to various special interest groups.) Because conservation has a net negative cost over the long term, it makes sense to give it the highest priority in any future revisions of national and international energy policy.

DOWNSIDE: Long-term, virtually none.

STOP USING COAL

Phase out the use of coal as a fuel. The state of Florida is already doing this. Despite promises that "clean coal" technology is on the horizon, it is inherently a dirty and bulky substance that creates pollutants at every step along the way—from mining it, to transporting it, to burning it, to disposing of the ash.

Sequestering coal's greenhouse emissions would require expensive investments. Surely it is better to spend that money on alternative methods of generating electricity than to build new coal-fired power plants.

DOWNSIDE: Most of the coal companies will suffer economically, as will their current employees as the transition takes place.

THE NUCLEAR OPTION

Make more use of nuclear power—particularly in the United States—at least as an interim solution. Amazingly, there turns out to be considerably more uranium on Earth than we knew about just a few decades ago.

Such a strategy, however, has numerous cons as well as pros. One major snag is waste disposal. The French minimize this problem by using breeder reactors, which permit them to reprocess their spent fuel as many as ten times. This not only saves money, but it also reduces the final quantity of nuclear waste to manageable amounts.

Nuclear reprocessing, however, also raises the spectre of nuclear materials getting into the hands of prospective terrorists.

DOWNSIDES:

1) Nuclear power creates a legacy of radioactive waste for future generations who will need to keep it contained for a millennium or more.

2) If developed countries expand their nuclear facilities, by what logic can they expect less-developed countries to voluntarily forego developing nuclear power?

WORST DOWNSIDE: A potential major disaster because nuclear materials get into the wrong hands.

UPSIDE: Nuclear power produces no greenhouse gases other than the small amounts generated in transporting the small volumes of fuel.

USE MORE RENEWABLE ENERGY

Be more aggressive about generating power from renewable resources: solar, geothermal, wind, tides, and ocean currents (the Gulf Stream, for instance). This includes addressing the related energy storage problems, since winds and tides cannot be started and stopped on demand. It is true that the initial capital outlays are high, because these energy sources are not highly concentrated. On the other hand, when such facilities are well designed, the operating costs are low, as no fossil fuels are being consumed. Recent studies suggest that the entire world could be powered by renewable sources within 30 years.

DOWNSIDE: The coal, oil, and some sectors of the transportation industries will take an economic hit.

UPSIDE: This initiative alone will create hundreds of thousands of new jobs. Moreover, in general, they will be *clean* jobs.

RENOVATE THE GRID

Upgrade obsolete power transmission grids. Currently, in the United States, for example, the electric power grid is in a sorry state. With the prospect of major brownouts and blackouts becoming ever more common, some sort of action is needed come what may. Such transmission upgrades need to take into account the prospective locations of future wind farms, geothermal facilities, and wave and tide generators.

DOWNSIDE: The money needs to come from somewhere.

UPSIDE: Once it is done, the improved grids should last a century or more.

RAIL TRAVEL

Expanding passenger railroads saves energy use elsewhere. A number of world cities already have light-rail commuter systems. Europe and Japan have remarkably efficient, high-speed inter-city rail systems (the Shinkansen "bullet train", the TGV, etc.) and probably are already ahead of the game in this regard. Yet the United States—the world's second-greatest greenhouse gas emitter—has virtually no inter-city passenger rail service these days. Even diesel-powered trains create (per passenger-mile) only a fraction of the emissions of aircraft or motor vehicles. If passenger trains became part of a nationwide energy-saving program, many of them would run on electricity generated from renewable sources.

DOWNSIDES:

1) Woes for the automobile industry, the airline industry, and (eventually) the oil companies.

2) The initial capital outlay will need to come from somewhere.

UPSIDE: Most of the rights-of-way already exist and are presently underused or are not used at all.

ELECTRIC VEHICLES

Continue to develop improved hybrid and electric vehicles. Continue research into advance battery technology.

DOWNSIDE: At present, such vehicles cost more than internal-combustion vehicles and their ranges are more limited.

UPSIDE: Their operating costs per mile are much less than petroleum-driven vehicles. Even if electricity begins to cost more, the operating cost of an electric vehicle will probably continue to be a bargain.

HYDROGEN AS FUEL

Continue the investigation of hydrogen (which can be extracted electrolytically from seawater) as an alternative fuel.

DOWNSIDE: Hydrogen is not as energy-dense as gasoline.

UPSIDES:

1) Its only combustion product is water vapor (H_2O).

2) There are places where it could fit into the existing infrastructure, thereby replacing piped natural gas, for instance.

NUCLEAR FUSION

Continue the international research on thermonuclear fusion—the process that fuels the Sun, which has been exploited explosively in the hydrogen bomb. Although the progress has been slow over the past four decades, the prospects are tremendous: Earth's oceans could fuel all of mankind's energy needs for tens of thousands of years into the future.

DOWNSIDE: Paying for the continued research.

UPSIDE: Gaining a great deal of collateral knowledge from that research.

NEW FOODSTUFFS

Develop acceptable alternatives to meat as a human food. Animal raising is the major anthropogenic contributor of methane to the atmosphere.

DOWNSIDES:

1) Changing people's tastes will probably take a generation or more.

2) Many sectors of the food industry will oppose the idea.

UPSIDE: Alternative sources of protein will probably be cheaper to the consumer than meat, poultry, and/or fish.

BURYING CARBON

There are proposals for removing the CO_2 from current smoke-stack gases, liquefying it, and pumping it into deep ocean trenches. There, the extreme pressure will keep it sequestered forever as a liquid. This is a brute-force approach to just one aspect of the problem, which would be quite expensive to implement.

In some places, it will be totally impractical. For emissions from large existing coastal power plants, however, this could be an interim solution—one that could allow those facilities to continue to operate for their full design lifetimes (typically about 40 years) before they are replaced by power plants that are more environmentally friendly.

DOWNSIDE: This added step would increase many people's electric bills.

UPSIDE: The added costs might be an incentive to further energy conservation.

GEOENGINEERING

Fund research on the most promising of the bevy of proposed geoengineering solutions to the greenhouse gas problem. These might include seeding sections of the southern oceans with powdered iron to stimulate blooms of plankton which absorb CO_2; increasing Earth's albedo by creating artificial clouds or other parasols; or bioengineering more efficient carbon-eating shrubs.

> DOWNSIDE: The history of technology is replete with stories of unanticipated negative consequences, and of good intentions gone awry.

> UPSIDE: Creating a repository of knowledge and strategies, which can be drawn upon, if and when necessary, in the future.

WHAT IF... ?

All of what I have tried to say before in these pages hardly completes the story of global climate change. New scientific results are published or otherwise disseminated almost daily. Some of these fresh findings will, ultimately, be replicated and confirmed, while others may ultimately be discredited or modified. Such details are really well beyond the scope of a book like this. What I *have* attempted instead is to focus on how the global climate issue arose in the first place, the facts that we are pretty sure of, how we have come to know those facts, and why they are so very important.

However, what if today's climate scientists have been too pessimistic in their projections? After all, even the IPCC places the odds of that at about one in 20. What if whole world panics, invests in major greenhouse gas mitigation programs, and then learns that the climatologists have, in the end, actually screwed up in predicting the imminence of global climate change? What then?

Would not nations around the globe have undertaken a lot of work for nothing? I think not. A tremendous amount of human capital is expended currently in drilling for petroleum, shipping it from one place to another, fighting wars because of it, attempting to prevent oil revenues from funding terrorists, and so on. The most valuable uses for oil are for making plastics, paints, solvents, pharmaceuticals, and the like—not to simply burn it away.

As for coal, it also provides raw materials for the chemical industry, but to burn it for energy requires huge amounts, and there are better and cleaner ways to get that energy (even oil is cleaner than coal).

Undoubtedly, coal and oil will continue to play roles in any conceivable future global economy, but not necessarily on the outlandish scales of modern-day consumption. Cutting back on that consumption, in other words, is justifiable even if the climate crisis turns out to be more benign than the current evidence warns.

Meanwhile, would the world's nations have spent a lot of money for very little return? Again, I think not; for whatever comes out of our response to the global climate issue, the scientific knowledge will last as long as humans walk the Earth. Moreover, all nations will have left their children, and their children's children, a far better planet on which to live.

Afterword

On the evening of April 20, 2010, 42 nautical miles south of the mouth of the Mississippi River, the $560-million offshore drilling rig Deepwater Horizon was rocked by a pair of violent explosions. Crude oil and flammable gas, relieved of the hydrostatic pressure that for eons had sequestered it deep beneath the seafloor, gushed to the drill deck and engulfed the giant platform in towering flames. A service ship, which at that time was preparing to offload drilling fluid, or "mud," rescued 115 survivors. Eleven others perished. Within 36 hours, the floating city of equipment and living quarters that had been the pride of Transocean Ltd. sank to the bottom in 5,067 feet (16,624 m) of water. Months later, crude oil from 13,293 feet (43,611 m) below the seabed continued to gush into the Gulf of Mexico from the rig's damaged wellhead.

That well, of course, was being drilled by British Petroleum, under a lease granted by the U.S. Department of the Interior and a permit approved by the U.S. Minerals Management Service. British Petroleum had assured the U.S. government that it was prepared to deal with a "worst-case scenario"—which according to its own documents would be a leak of 160,000 barrels per day. The company promised that little, if any, of such a spill would ever reach the coastlines.

In fact, the actual spill rate was considerably lower—about 40,000 barrels (1.7 million gallons) per day—although estimates did vary from time to time and the rate itself probably fluctuated. BP, however, proved to be incapable of handling even a fraction of this figure. The slick spread over several thousand square miles, despite the efforts of some 2,000 vessels to scoop it up. Huge plumes of oil also blossomed beneath the surface. And the goo oozed into at least 500 miles of Gulf coastline—most notably the fragile wetlands of southern Louisiana. The event quickly became one of the most difficult and damaging environmental disasters in history.

There had, however, been big oil spills before. In 1969, there was one in California's Santa Barbara Channel, which polluted 35 miles of shoreline, killed thousands of birds and even some seals and dolphins, and took several years to clean up completely. In 1979, in Mexico's Bay of Campeche, a drilling rig exploded and caught fire, and over the next nine months about 72,000 barrels of oil washed up on 162 miles of Texas beaches. And in 1989, the supertanker *Exxon Valdez* ran aground in the remote Prince William Sound in Alaska, spilling about 250,000 barrels of crude oil and gunking up 1,300 miles of coastline. Today, more than 20 years later, the courts still have not resolved all of the legal claims resulting from that particular incident.

In other words, disasters of this type seem to occur with some regularity at 10-year intervals—just long enough for everyone to get lulled into a sense of complacency once again, and to forget the passel of risks associated with the world's continued dependence on dirty sources of energy.

As this book goes to press, the Deepwater Horizon disaster remains far from over. The cleanup alone will take a year or more, and the recovery of the marshes (those that do recover rather than disappearing forever) will take several years. Then, if history is any guide, the lawsuits will drag on for another decade or more.

Will the ultimate outcomes be worth it?

British Petroleum—the fourth most profitable company in the world—will certainly survive. Losing a few tens of billions of dollars won't hurt it much in the long run. And the pristine Gulf beaches will be restored fairly quickly. It might be expensive, but it isn't difficult to scoop up oily sand and haul it away to be incinerated.

The marsh grasses that have been smothered, however, will decay and the fragile land they hold in place will erode. Shrimp and many species of fish will lose their spawning grounds. Offshore, microbes eating the uncollectible emulsified oil will deplete the oxygen in large swaths of the Gulf, creating huge new "dead zones" where fish and other sea creatures cannot survive. Much of the seafood that *is* harvested will be found to contain unacceptably high levels of toxins. And many commercial fishermen will lose their source of livelihood.

Meanwhile, most of the roughly 4,000 active offshore wells (the precise census actually changes from month to month) will continue to extract oil from beneath the Gulf of Mexico without creating a disaster. That oil will continue to be refined, and the resulting gasoline, diesel fuel, and jet fuel will continue to be burned in a variety of internal combustion engines. And, yes, carbon dioxide will continue to enter Earth's atmosphere.

* * *

An oil spill is an acute disaster. It is localized in time and space, and it's highly visible. It immediately grabs everyone's attention regardless of their political views.

The buildup of CO_2 in the atmosphere is considerably more subtle. Earth has just one atmosphere, so the effects are global rather than regional. It's also a slow process compared to an oil spill, with the atmospheric composition changing by only a few parts-per-million per year. And finally, you can't see CO_2—unless you have the proper instruments and you're paying attention to them.

The two disasters, however, are inextricably linked. If we weren't burning fossil fuels, we would have neither a climate-change issue nor oil spills.

No, I'm not saying we should abruptly halt our use of petroleum. We might, however, choose to begin suppressing our appetite for it. Although fueling aircraft with anything other than hydrocarbon-based fuels remains a huge technological challenge, cars powered by electricity are now commercially practical. The economic comparison: a gasoline-powered car driven 15,000 miles per year and getting 30 miles per gallon has an annual fuel cost (in the U.S.) of about $1,500. A similar car using electricity at 12 cents per kilowatt-hour has an annual energy cost of about $1,100.

And how would the necessary electricity be produced, if not from fossil fuels? There are clean and economical ways to do that, too—using wind or solar cells. Wind turbines, for instance, presently cost about $1 million per megawatt of generating capacity. Suppose that the price tag of the Deepwater Horizon disaster—which will ultimately add up to at least $4 billion if not more—had been invested instead in offshore wind turbines. That single wind farm alone (even if it ran at only 50 percent of capacity) would produce enough electricity to power about 2 million electric cars. At that rate, the initial capital investment would be recovered in less than three years.

Yes, the economics are there. And yes, the technology is available.

But is the political courage?

INDEX